AF469689

Guide to Acceptance Sampling

Dr. Wayne A. Taylor

Taylor Enterprises, Inc.
Lake Villa, Illinois

ISBN   0-9635122-0-X

Printed in the United States of America by
R. R. Donnelley & Sons

Cover designed by
Ann B. Taylor

Cover artwork prepared by
Sue Wuerffel

Edited by
Harold Bailey
Ann B. Taylor

Current Printing (Last Digit): 10 9 8 7 6 5 4 3 2 1

To Ann, Diana, and Lisa

Thanks to all those that have aided me in this effort starting with my wife Ann. Ann designed the cover and aided in the editing. She also provided me with the time needed to complete this book. I would also like to thank Harold Bailey for his careful proofing of the book and his continuing support in many other ways. Credit goes to Sue Wuerffel for taking the cover design and translating it into a printed cover. I would also like to thank Baxter Healthcare Corporation for its support of my writing efforts and give particular thanks to Ron Abrahams, Beth Dudley, Harold Sargent, Gerry Bordash, Dave Ingram, and Bob Machalski to name but a few. Your comments, suggestions and support are most appreciated. Bob designed the checker charts originally used for QSSs and which appear in this book in modified form. Finally, I would like to thank Jack Doyle for sparking my original interest in quality control and acceptance sampling.

Contents

Tables

Introduction

The objectives of this book are to

- Improve the effectiveness of acceptance sampling.

- Improve the efficiency of acceptance sampling: Alternate types of sampling plans can provide equivalent protection at reduced costs.

Despite being one of Quality's oldest tools, acceptance sampling is commonly misunderstood and frequently misused. These misconceptions and harmful practices waste valuable resources and undermine the effectiveness of acceptance sampling. Improving the effectiveness and efficiency of acceptance sampling begins with building a better understanding.

This book teaches the different uses of acceptance sampling. It also teaches how to select and evaluate acceptance sampling plans for these different uses. It is written for the practitioner. All needed aids are included. Complex mathematical formulas are avoided. Instead, tables and computer programs are used to simplify the tasks. A disk is included containing several programs that run on an IBM® PC or compatible. Versions for other computers are available.

This book goes well beyond the selection and evaluation of sampling plans. It explores the proper usage of acceptance sampling. Alternatives to acceptance sampling such as 100% inspection and no inspection are also explored. Procedures are given for selecting the best course of action. Further, the practical aspects of acceptance sampling are covered including the formation of lots and the selection of samples. Finally, guidance is provided for combining acceptance sampling with SPC and process monitoring in order to avoid redundancy and improve cooperation.

Special attention is given to selecting efficient sampling plans that reduce costs. This book contains two important advancements not covered in earlier books. First, the quick switching systems in Chapter 7 can be used to reduce sampling costs by as much as 85% for continuous processes. Second, the procedure for selecting double sampling plans in Chapter 6 results in double sampling plans that are more efficient than those found in Mil-Std-105E and elsewhere. Together, these advancements can significantly reduce sampling costs: typically by 20% to 30%. For large corporations this can amount to millions of dollars a year.

Acceptance sampling can be applied under a variety of circumstances. As such, there are many different types of acceptance sampling plans. This book is limited to acceptance sampling plans for controlling the rate of defects or defectives. Not covered are sampling plans for controlling the MTBF (mean time between failure), the average, and the standard deviation. This book also omits many less frequently used sampling methods. It concentrates on those methods making up the backbone of acceptance sampling. Covered are single sampling plans, double sampling plans, quick switching systems and variables sampling plans. In addition, Mil-Std-105E is covered in depth.

1.1   Definition of Acceptance Sampling

Acceptance sampling is used to decide whether to accept or reject lots of product. It is performed once the lots are produced. Rejected lots might be discarded, 100% inspected, or sold at a discount. The basic procedure consists of forming a lot of product, selecting a sample from the lot, inspecting the units in the sample, and using the results to decide whether to accept or reject the lot. Such sampling plans can be used to inspect incoming materials, in-process components and finished goods.

Consider the inspection of apples to be used in the making of applesauce. Shipments where more than 2% of the apples are bruised or spoiled will result in substandard applesauce. A sampling plan is used to decide which shipments are usable. Each truckload of apples is considered a lot. A sample of fifty apples is selected from each lot and inspected for bruised or spoiled apples. Those lots with samples containing one or fewer bad apples are accepted for use.

Acceptance sampling is one type of sampling inspection. Sampling inspection involves the selection of samples from a process and the use of these samples to make decisions. Acceptance sampling decides whether to accept or reject lots of product. One can also use samples to decide whether to adjust the process. This is called process control. Samples can also be used for other purposes such as estimating the process average (See Figure 1.1). The same samples used for acceptance sampling are often used for these other purposes also. Therefore, it is important to select acceptance sampling plans that fit with these other purposes.

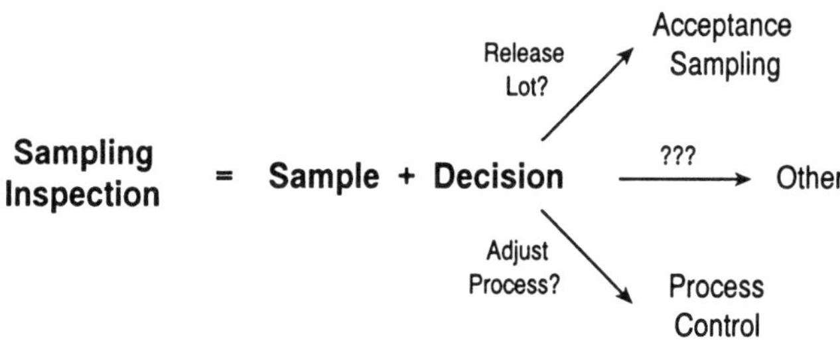

Figure 1.1: Sampling Inspection

1.2   Purpose of Acceptance Sampling

The purpose of acceptance sampling is to improve the quality of the product going to the customer. It does this directly by rejecting lots. Sampling plans for this purpose can be selected to

- Insure against the release of highly defective lots.

- Insure the average quality going to the customer is better than some specified level.

- Improve the quality of the product going to the customer in a cost effective manner so as to maximize the customer value.

- Reject obviously bad lots.

Acceptance sampling also improves quality indirectly by providing incentives for improving quality and penalties for deteriorating quality. Sampling plans are frequently selected with more than one of these purposes in mind.

What acceptance sampling cannot do is provide 100% perfect product. The only way of doing this is to <u>make</u> 100% perfect product.

Acceptance sampling is not an alternative to process improvement. Instead, acceptance sampling should complement improvement efforts by providing incentives to improve. It also serves as a temporary measure for ensuring customer quality until successful improvements are made.

Once improvements have been made, one cannot simply walk away from the process. No matter how long the process has been running without experiencing a problem, there is still the possibility that something will go wrong in the future. Suppliers may change, materials may vary and equipment can break. Here, acceptance sampling can serve as an insurance policy against the release of highly defective lots.

Even if one decides not to use acceptance sampling, some samples should be periodically inspected to monitor process performance. Samples may also be required to control the process. Whenever samples are taken, no matter what the purpose, one should always have a procedure in place for rejecting obviously bad lots. Release of a lot where samples exist clearly indicating the lot is bad creates a significant liability. This book provides procedures for selecting sampling plans for each of the above purposes.

1.3   Alternatives to Acceptance Sampling

The two alternatives to acceptance sampling are 100% inspection and 0% inspection. In reality, one has a choice of five different actions:

- Release all lots without any inspection

- Discard all lots without any inspection

- 100% inspect all lots

- Acceptance sample with rejected lots being discarded

- Acceptance sample with 100% inspection of rejected lots

There are several minor variations. Discarded lots might be sold at a discount. Defective units found as part of 100% inspection can be discarded, repaired or replaced. If testing is destruction, those actions involving 100% inspection are not possible.

Acceptance sampling is not of benefit unless lot quality varies. If all lots are of the same quality, all lots should receive the same treatment: either 100% inspection or no inspection. When lot quality varies, the best choice depends on the costs involved and the quality of the lots. Acceptance sampling generally proves advantageous when the cost of inspection is high or when the 100% inspection is not very reliable. One hundred percent inspection generally proves advantageous for critical defects or when the cost of inspection is small. Selecting the best action requires careful analysis. Chapter 4 covers this decision in detail.

1.4   Variety of Situations

Acceptance sampling is applied under a variety of situations. Inspection results might be reported on a pass/fail basis or as measurements. The lot to be tested could be an isolated lot or one of a series. Testing could be destructive or nondestructive. A list of the many possibilities is given in Table 1.1.

**Table 1.1: Different Situations Under Which
Acceptance Sampling Is Applied**

Category	Different Situations
Type of Data	pass/fail, counts of defects, measurements
Destructive Testing	destructive, nondestructive
Group Testing Possible	test as a group, test one at a time
Cost of Testing	high, medium, low
Type of Lot(s)	isolated, sequence
Point of Inspection	receiving, in-process, final
Disposition of Rejected Lots	discard, discount, 100% inspect
Disposition of Defectives	discard, replace, rework

Because of the variety of situations under which acceptance sampling is applied, there are many different types of sampling plans. This book covers single sampling plans, double sampling plans, quick switching systems and variables sampling plans. None of these types is best suited to all situations. This book will show you how to select sampling plans of each of these types, compare them, and then select the best sampling plan for the situation.

1.5   Summary

Acceptance sampling is a type of sampling inspection that decides whether to accept or reject lots of product. Samples used for acceptance sampling are frequently used for other purposes as well. This can affect the acceptance sampling plan selected.

The purpose of acceptance sampling is to improve the quality going to the customer. It does this directly by improving the outgoing quality and indirectly by providing incentives for improving the incoming quality. Acceptance sampling cannot ensure 100% perfect product. It is not an alternative to process improvement.

Alternates to acceptance sampling are 100% inspection and 0% inspection. In reality, one must choose between releasing all lots, discarding all lots, 100% inspecting all lots, acceptance sample with rejected lots being discarded, and acceptance sample with 100% inspection of rejected lots. The best choice depends on whether testing is destructive, the costs involved, and the quality of the lots. Many different types of sampling plans exist due to the variety of situations under which acceptance sampling is applied.

Further Basic Reading

> Duncan, Acheson J. (1986). <u>Quality Control and Industrial Statistics</u>. Fifth Edition, Richard D. Irwin, Homewood, Illinois.

> Grant, Eugene L. and Leavenworth, Richard S. (1988). <u>Statistical Quality Control</u>. Sixth Edition, McGraw-Hill, New York, New York.

Further Advanced Reading

> Schilling, Edward G. (1982). <u>Acceptance Sampling in Quality Control</u>. Marcel Dekker, New York, New York.

2

Evaluating
Single Sampling Plans

The objectives of this chapter are to

- Learn to make accept and reject decisions using single sampling plans.

- Learn to evaluate the protection provided by single sampling plans.

It is assumed that a single sampling plan has already been selected. Chapters 3 and 4 show how to select single sampling plans. Other types of sampling plans will be explored in the remaining chapters.

2.1   Characterizing Single Sampling Plans

Single sampling plans have two parameters:

n = sample size

a = accept number

The sample size is the amount of product to inspect. In most cases, it is the number of units to inspect. However, the sample size can also be a quantity such as 3.5 square feet of plastic sheeting or half a cubic yard of cement. The number of defects or defective units in the sample is tallied. Those lots whose tally exceeds the accept number are rejected. Figure 2.1 shows the procedure used by single sampling plans in deciding which lots to accept and reject.

Chapter 1 gives an example involving the inspection of truckloads of apples. The single sampling plan n=50 and a=1 is used. For each

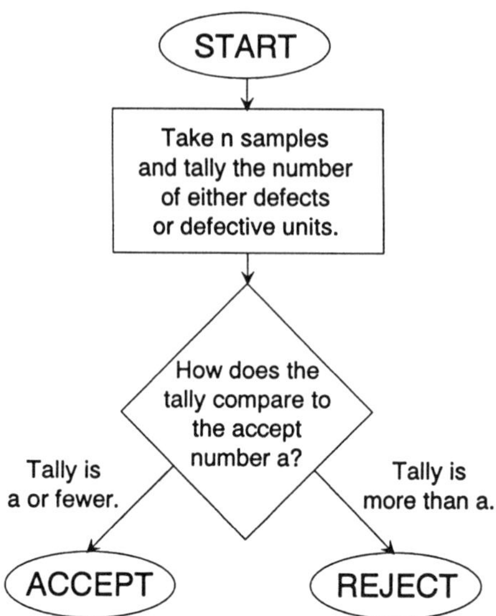

Figure 2.1: Procedure for Single Sampling Plans

truckload, a sample of fifty apples is selected, and the number of bruised or spoiled apples tallied. Table 2.1 shows the results of inspecting 10 truckloads. Lots whose sample contains one or fewer bad apples are accepted; Two or more bad apples results in rejection.

Table 2.1: Inspection Results

Q. C. Checker Chart			
Product: <u>Apples</u>	Quantity: <u>Truckload</u>	Sample Size: <u>50</u>	Accept Number: <u>1</u>
Truckload	**Number Defectives**	**Decision**	**Initials**
91-07-1	0	accept	WT
91-07-2	0	accept	WT
91-07-3	5	reject	WT
91-07-4	0	accept	WT
91-07-5	2	reject	WT
91-07-6	0	accept	HB
91-07-7	0	accept	HB
91-07-8	1	accept	HB
91-07-9	0	accept	HB
91-07-10	0	accept	HB

When using single sampling plans, one has the choice of tallying either the number of defective units or the total number of defects. A defect is the lack of a desirable characteristic or the presence of an undesirable characteristic. A defective unit is any unit containing one or more defects. Some units can contain more than one defect. When sampling a quantity such as half a cubic foot of cement, one can only tally the number of defects. When sampling units of product, one can also tally the number of defective units.

When the defect rate is low, rarely will a single unit contain more than one defect. In this case it makes little difference whether one tallies the number of defects or the number of defective units. It is merely a matter of convenience. For higher defect rates, valuable information is lost when the number of defective units is tallied. In this case, the number of defects should be tallied.

Some prefer to use the terms nonconformance and nonconforming unit in place of defect and defective unit respectively. These terms emphasize that failure to meet the specification does not necessarily imply that the unit is unfit for use. Some specifications reflect process requirements and have no effect on customer quality. This book uses the wording defect and defective unit. However, any time these two terms are used, the terms nonconformance and nonconforming unit can be substituted.

Exercises

(2.1)   The single sampling plan n=13 and a=0 is used to inspect incoming cartons of machine screws. One carton's sample contains one defective screw. Should this carton be accepted or rejected? The next carton's sample contains no defective screws. Should it be accepted or rejected?

(2.2)   Plastic sheeting is being inspected for holes. Every hour a sample of 2.5 square yards is inspected. Should one tally the number of defects or the number of defective units? Why?

(2.3)   Suppose one is inspecting for bruised or spoiled apples. Typically 1% are bad. Should one tally the number of defects or defectives? Why?

(2.4)   When inspecting an airplane wing for damaged rivets, it is not uncommon to find from 5 to 10 bad rivets. Should one tally the number of defects (bad rivets) or the number of defectives (wings with one or more bad rivets)? Why?

2.2   Probability of Acceptance

To investigate the protection provided by single sampling plans, let us examine the behavior of the single sampling plan n=50 and a=1. If this sampling plan is used to inspect a series of lots from a 1% defective process, some samples will contain no defectives, some will contain one defective and so on. Lots whose sample contains one or fewer defectives are accepted. All other lots are accepted. Some lots will be accepted and some will be rejected. If the process is running 1% defective, what is the probability of a lot being accepted?

To find out, a 1% defective process was inspected 100 times. The results are shown in Figure 2.2. Of the 100 samples, 61 contained no defectives, 33 contained one defective and so forth. Of the 100 inspections, 94% resulted in accept decisions.

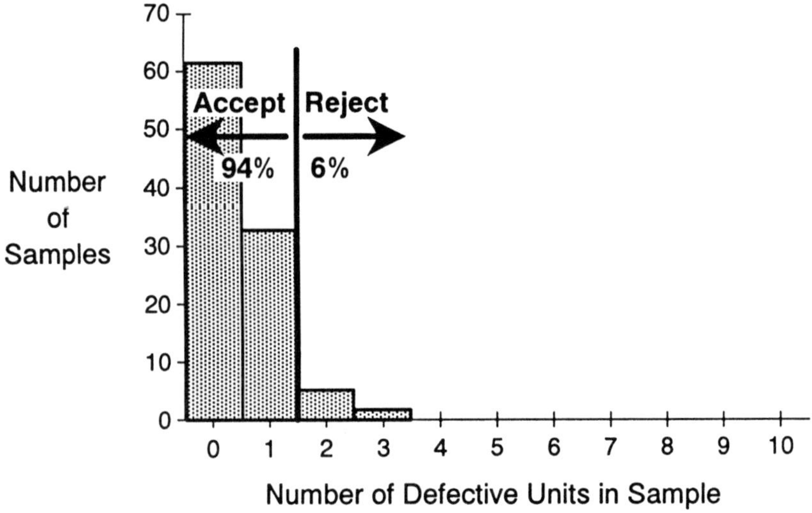

Figure 2.2: Results of 100 Inspections of a 1% Defective Process Using the Single Sampling Plan n=50 and a=1

Suppose the process quality gets worse. Figure 2.3 shows the results of 100 inspections performed on a 4% defective process. Of these inspections, 46% resulted in accept decisions. Figure 2.4 shows similar results for a process running 10% defective. In this case, 7% of the inspections resulted in accept decisions.

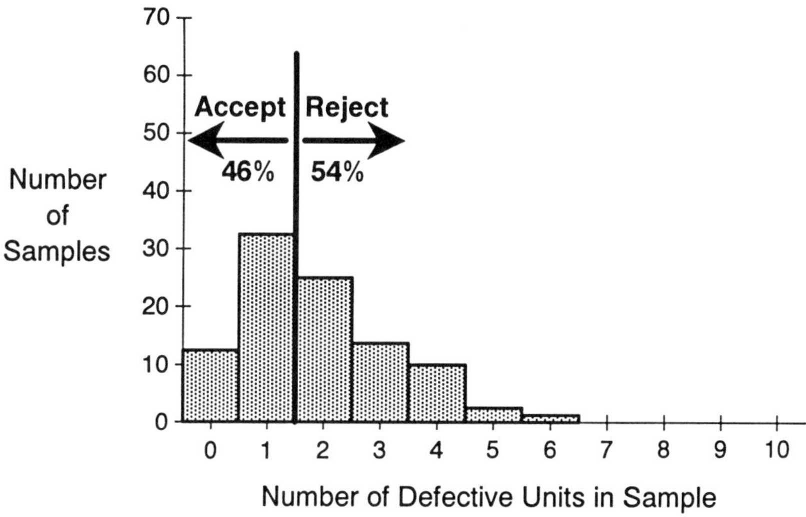

Figure 2.3: Results of 100 Inspections of a 4% Defective Process Using the Single Sampling Plan n=50 and a=1

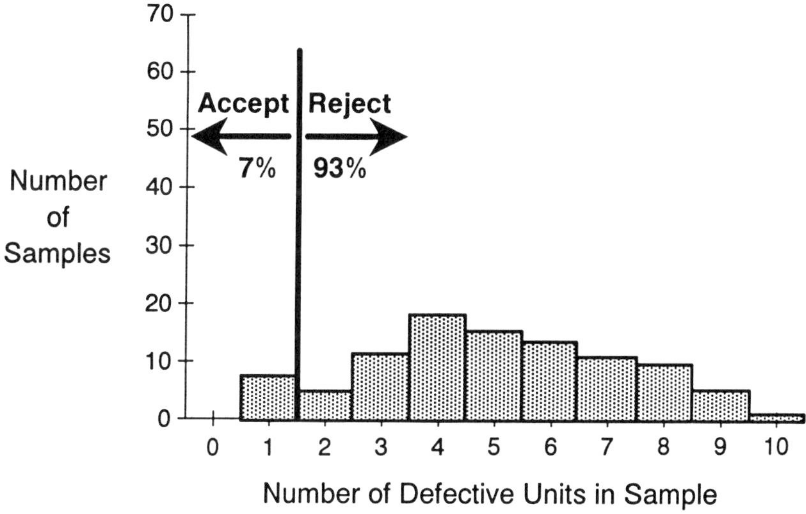

Figure 2.4: Results of 100 Inspections of a 10% Defective Process Using the Single Sampling Plan n=50 and a=1

The behavior of the single sampling plan n=50 and a=1 can be summarized as follows:

- 1% defective processes are generally accepted (~94%).

- 4% defective processes are accepted half of the time (~46%).

- 10% defective processes are rarely accepted (~7%).

Rather than simulate the performance of a sampling plan, it is possible to calculate the probability of acceptance. Formulas are given in Appendix B. These formulas are fairly complex. However, it will not be necessary to use these formulas. The enclosed software performs all the required calculations. Appendix B is included to provide a greater level of detail for those who desire it; and to serve as documentation of the calculations performed by the enclosed software.

Being able to calculate the probability of acceptance allows one to plot process quality versus the corresponding probability of acceptance. Such a plot is called an operating characteristic curve or OC curve for short. The OC curve of the single sampling plan n=50 and a=1 is shown in Figure 2.5. The bottom axis gives the process quality in terms of percent defective. The left axis gives the corresponding probability of acceptance. From this OC curve we see that 1% defective lots are accepted 91% of the time, 4% defective lots are accepted 40% of the time, and 10% defective lots are accepted 3% of the time. The simulation results are in close agreement with these values. OC curves are the topic of the next section.

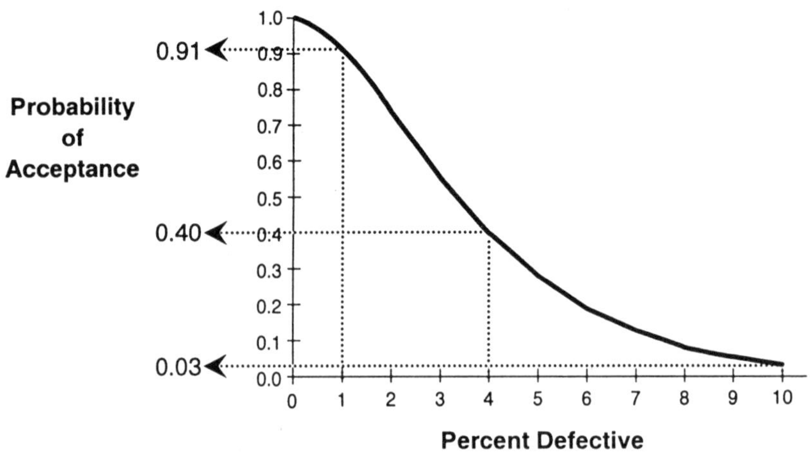

Figure 2.5: OC Curve of Single Sampling Plan n=50 and a=1

2.3   OC Curves

An OC curve is a plot of process or lot quality versus the probability of acceptance. Figure 2.5 gave an example. Each sampling plan has its own characteristic OC curve. This OC curve summarizes the protection provided by the sampling plan. From the OC curve, one can determine what quality lots are routinely released and routinely rejected. All OC Curves have certain properties in common:

- At 0% defective the probability of acceptance is 1.

- At 100% defective, the probability of acceptance is 0.

- As the percent defective increases, the OC curve decreases.

However, there can be great variation in the steepness of the OC curve and the point at which the OC curve starts to decline.

Ideally, sampling plans should release good quality lots and reject poor quality lots. If lots better than 0.5% defective are considered good, the ideal OC Curve would appear as in Figure 2.6.

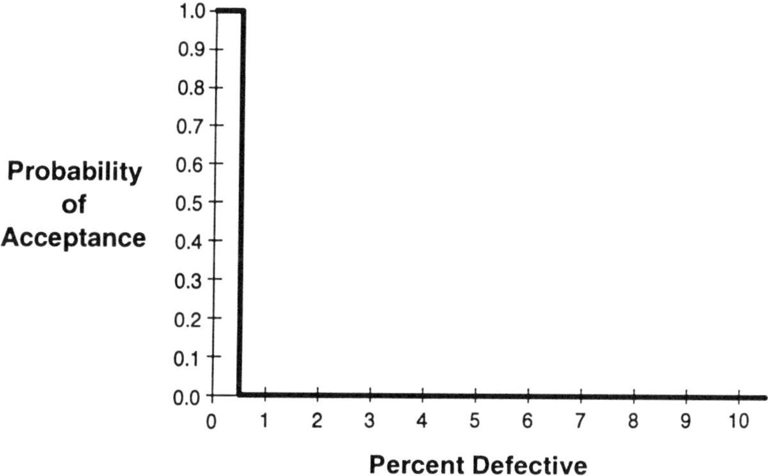

Figure 2.6: Ideal OC Curve

This ideal can only be obtained by inspecting the entire lot. This can be prohibitively expensive. Because sampling plans do not examine the entire lot, they run the risk of reaching wrong decisions. Compromises must be made to balance the number of units inspected against the chances of making errors. Figure 2.7 shows several OC curves that might be used instead. The far left OC curve ensures bad lots are rejected. It does so at the risk of rejecting good lots. The far right OC curve ensures good lots are released. It also accepts some bad lots. The middle OC curve balances the chances of making these two types of errors.

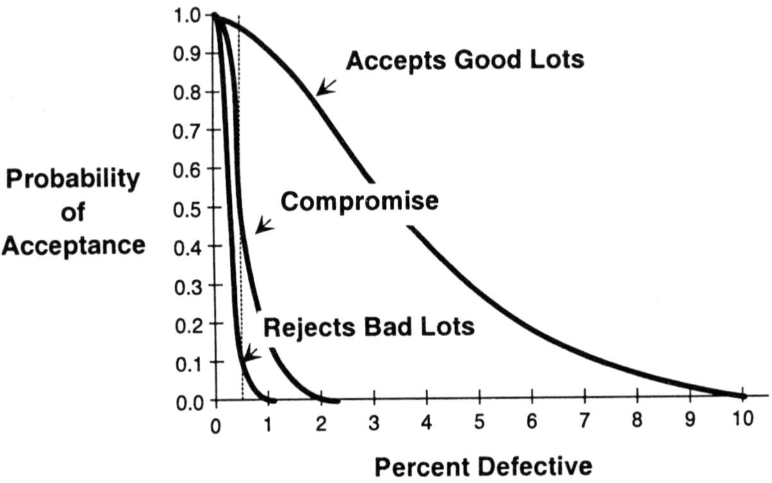

Figure 2.7: Different OC Curves

Every single sampling plan has four different OC curves. They are:

- Type-B OC Curve for Defectives
- Type-B OC Curve for Defects
- Type-A OC Curve for Defectives
- Type-A OC Curve for Defects

Figures 2.8 to 2.11 show the four different OC curves for the single sampling plan n=50 and a=1. The appropriate OC curve to use depends on whether the number of defects or defectives is tallied and on whether one is describing the protection for an individual lot or for a series of lots.

The Type-A and Type-B OC curves for defectives are used when one tallies the number of defectives. When defectives are tallied, the bottom axis gives the process or lot quality in percent defective. When defects are tallied, the Type-A and Type-B OC curves of defects should be used instead. In this case the process or lot quality on the bottom axis is in terms of the average number of defects per unit.

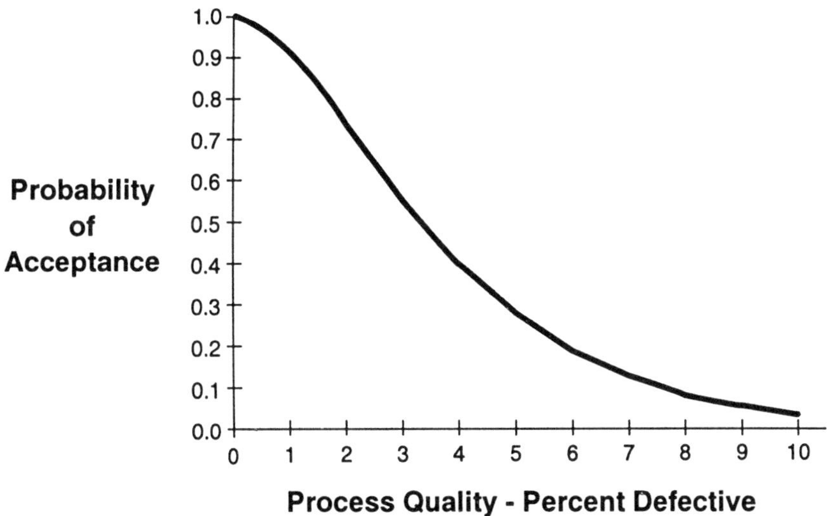

**Figure 2.8: Type-B OC Curve for Defectives
Single Sampling Plan n=50 and a=1**

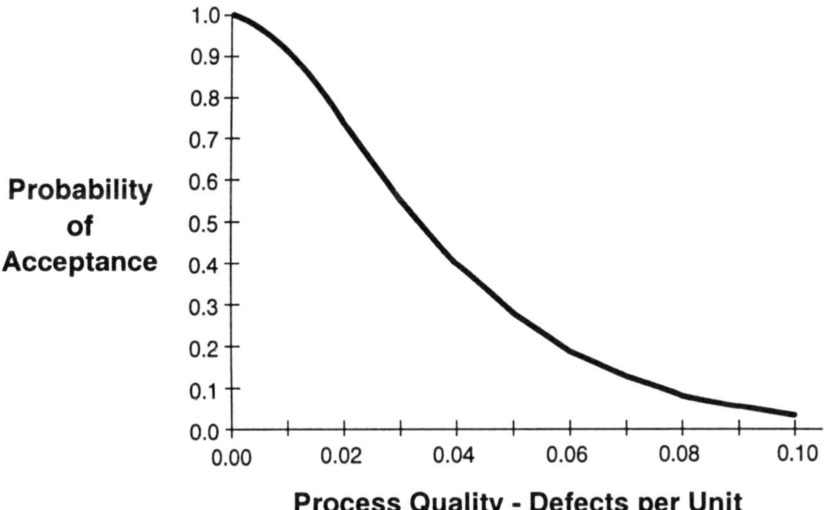

**Figure 2.9: Type-B OC Curve for Defects
Single Sampling Plan n=50 and a=1**

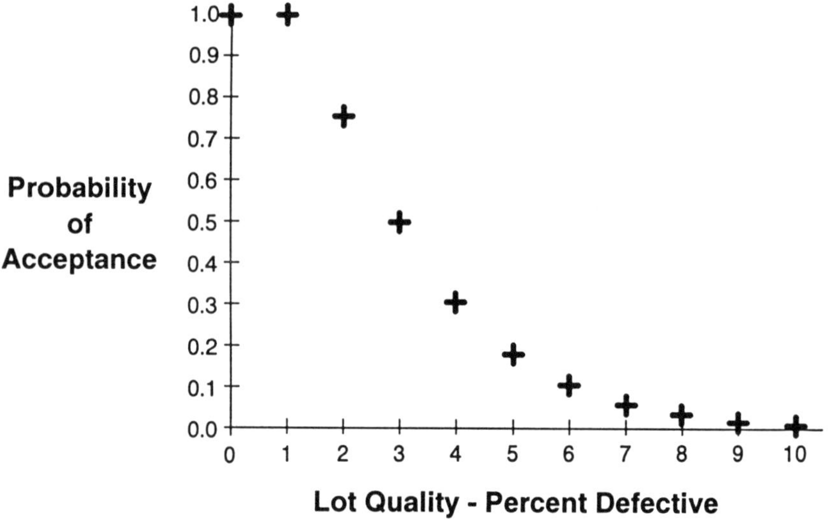

Figure 2.10: Type-A OC Curve for Defectives, Lot Size = 100
Single Sampling Plan n=50 and a=1

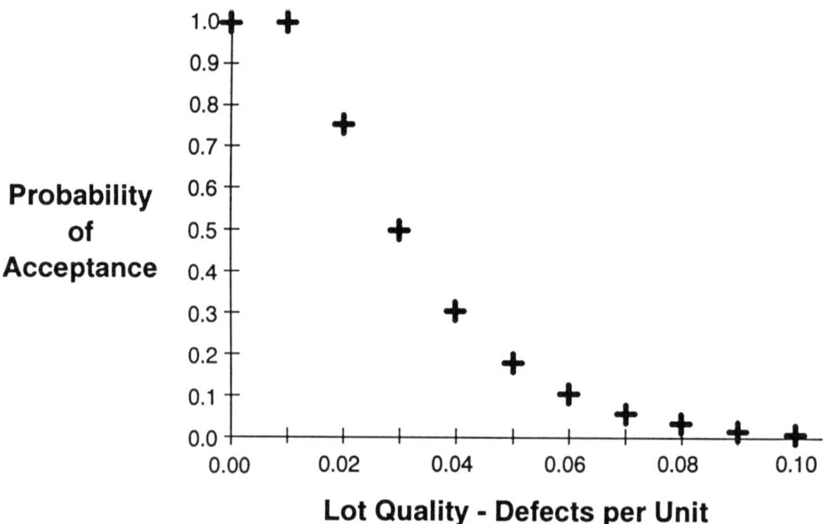

Figure 2.11: Type-A OC Curve for Defects, Lot Size = 100
Single Sampling Plan n=50 and a=1

Type-B OC curves describe the protection provided by a single sampling plan when applied to a series of lots from a process. The bottom axis gives the process quality, either as a percent defective or in terms of the average number of defects per unit. Type-B OC curves are smooth curves.

Type-A OC curves describe the protection provided by a sampling plan for an individual lot. The bottom axis gives the lot quality. Lot quality is expressed in percent defective or the average number of defects per unit. Lot quality is calculated as follows:

$$\text{Lot Percent Defective} \quad = \quad 100 \times \frac{X}{N}$$

$$\text{Average Number of Defects per Unit} \quad = \quad \frac{X}{N}$$

X is the number of defects or defectives in the lot and N is the lot size. Since X is an integer, lot quality is restricted to specific values. As a result, a Type-A OC curve is a series of points rather than a smooth curve. Table 2.2 summarizes the four types of OC curves.

Table 2.2: Four Types of OC Curves

		Tally?	
		Defectives	**Defects**
Protection?	**Individual Lot**	**Type A for Defectives** Lot Quality in Percent Defective versus Probability of Accepting Lot	**Type A for Defects** Lot Quality in Defects per Unit versus Probability of Accepting Lot
	Series of Lots from Process	**Type B for Defectives** Process Quality in Percent Defective versus Probability of Accepting Lots From Process	**Type B for Defects** Process Quality in Defects per Unit versus Probability of Accepting Lots From Process

OC curves of single sampling plans can be generated using the program SINGLE on the Sampling Programs diskette. Appendix A contains detailed instructions for running the software. The first step is to obtain the DOS prompt. The DOS prompt looks similar to: "C:\>". Once obtaining the prompt, place the Sampling Programs diskette in one of the disk drives. Then start the program by entering either **A:SINGLE** or **B:SINGLE** depending on whether the diskette is in the A or B drive. Entering **A:SINGLE** means to press the following sequence of keys: "A", ":", "S", "I", "N", "G", "L", "E" and "Enter". Starting the program results in Screen 2.1.

Screen 2.1: Main Menu (SINGLE)

```
***********************************************************************************
*                              *                                                 *
*                              *        Copyright (C) 1992 Taylor Enterprises    *
*     Program SINGLE           *         P.O. Box 820, Lake Villa, IL 60046      *
*                              *               (708) 356-1074                    *
*                              *                                                 *
***********************************************************************************

                   Current Sampling Plan:   no plan selected



                       -----  MENU OPTIONS  -----

             (1) Enter sampling plan
             (2) Summary information
             (3) OC curve
             (4) AOQ curve
             (5) ASN curve
             (6) Select sampling plan based on AQL and LTPD
             (7) Select sampling plan based on AQL and AOQL
             (8) Select sampling plan based on AQL and n


 ENTER NUMBER OF OPTION OR "q" TO QUIT   -->
```

Before an OC curve can be generated, the sampling plan must be entered. Option 1 is used to enter the sampling plan. Option 3 is then used to generate its OC curve. When entering the sampling plan, the program will ask whether defects or defectives are tallied and will request the sample size and accept number. After typing each response, press the enter key. If defects are tallied, the program will also ask whether the lot consists of units of product or is a continuous lot such as a tank of solution or load of cement. For continuous lots, non integer sample sizes are allowed.

Screen 2.2 shows the input of the single sampling plan n=50 and a=1 when defectives are tallied. The values entered by the user are preceded by the prompt: "-->". Once this information is entered, the program returns to the main menu with the single sampling plan displayed at the top.

Screen 2.2: Entering Sampling Plan (SINGLE)

```
***************************************************************************
***                                                                     ***
***              SINGLE Option 1 - Enter Single Sampling Plan           ***
***                                                                     ***
***************************************************************************

    Enter the sample size "n" and accept number "a" of the desired single
    sampling plan.  You must also specify whether defects or defectives
    are tallied.  If defects are tallied, one must further specify whether the
    lot consists of individual units or is a continuous lot such as a tank
    of solution or load of cement.  When defects are tallied and the lot is
    continuous, the sample size is not limited to integer values.

TALLY? (1=defectives,2=defects)            --> 1
ENTER SAMPLE SIZE    (n)                    --> 50
ENTER ACCEPT NUMBER (a)                     --> 1
```

Option 3 can now be selected to generate a plot or table of the OC
curve. Upon selecting option 3, Screen 2.3 appears asking for the
type of OC curve, whether a plot or table is desired, and how to scale
the plot or table. If a Type-A OC is requested, the lot size will also be
prompted for. In Screen 2.3, a plot of the-Type-B OC curve is
requested using the default scale. The resulting plot of the OC curve
is shown in Screen 2.4.

Screen 2.3: Entering Information for Plot of OC Curve (SINGLE)

```
***************************************************************************
***                                                                     ***
***                    SINGLE Option 3 - OC Curve                       ***
***                                                                     ***
***************************************************************************

    This option plots or tabulates the OC curve.  You must specify which type
    of OC curve to use:

         (A) Type A: describes protection for individual lots
         (B) Type B: describes protection for a series of lots, i.e., a process

    For Type-A OC curves, the lot size must be specified.  One may specify
    the scale or let the program select the appropriate scale.

ENTER TYPE OF OC CURVE. ("A" or "B")      --> b
ENTER "p" FOR PLOT "t" FOR TABLE.         --> p

    The default scale for percent defective (bottom axis) is from 0%
    to  10.0000000% defective.

USE DEFAULT SCALE? ("y" or "n")           --> y

    The default scale for probability of acceptance (left axis) is from 0 to 1.

USE DEFAULT SCALE? ("y" or "n")           --> y
```

Screen 2.4: Plot of OC Curve (SINGLE)

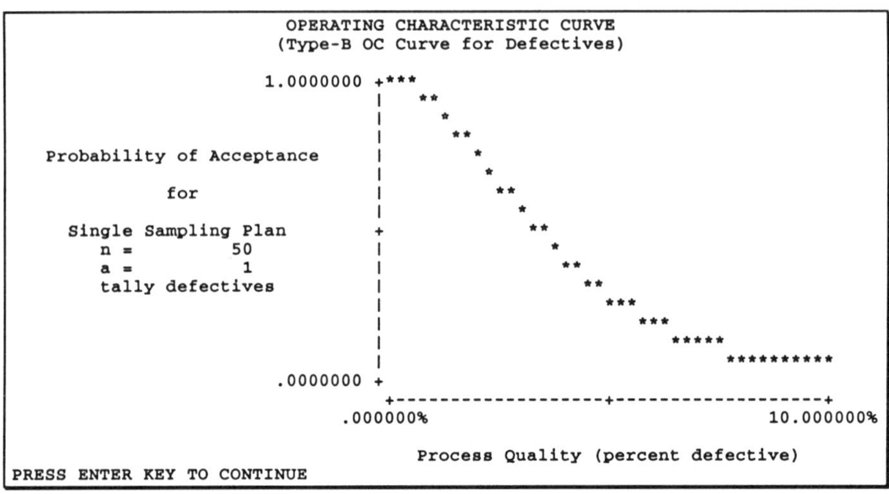

A printout of the OC curve can be obtained by holding down the "Shift" key and pressing the "Print Scrn" key. The "Shift"+"Print Scrn" keystroke combination can be used at any time to print the contents of the screen.

The OC curve can also be displayed in the form of a Table. Tables can be used to obtain accurate values of the probability of acceptance. Screen 2.5 shows the input required to generate a table of the Type-B OC curve using the default scale. The program suggests that the table start at 0% defective and go up to 10% defective using a step size of 1% defective. Using these values, the table will give the probability of acceptance when the process averages 0% defective, 1% defective, 2% defective, and so on up to 10% defective. Screen 2.6 shows the table generated using the recommended scale. If you want to use a different scale, enter "N" in response to the question: "Use default scale?" Then type the minimum and maximum percent defectives along with the step size.

Table 2.3 (page 22) gives probabilities of acceptance for the single sampling plan n=50 and a=1 when defectives are tallied. Values are given for Type-A OC curves with various lot sizes as well as for the Type-B OC curve. As the lot size gets large, the Type-A OC curves converge to the Type-B OC curve. As a rule of thumb, when the lot size is at least 10 times the sample size, there is little difference between Type-A and Type-B OC curves. In which case, Type-B OC curves can be used to approximate Type-A-OC curves. This is true regardless of whether defectives or defects are tallied.

Screen 2.5: Entering Information for Table of OC Curve (SINGLE)

```
******************************************************************************
***                                                                        ***
***                     SINGLE Option 3 - OC Curve                         ***
***                                                                        ***
******************************************************************************

   This option plots or tabulates the OC curve.  You must specify which type
   of OC curve to use:

       (A) Type A: describes protection for individual lots
       (B) Type B: describes protection for a series of lots, i.e., a process

   For Type-A OC curves, the lot size must be specified.  One may specify
   the scale or let the program select the appropriate scale.

ENTER TYPE OF OC CURVE. ("A" or "B")     --> b
ENTER "p" FOR PLOT "t" FOR TABLE.        --> t

   The default is for the table to start at 0% defective and go up
   to  10.0000000% defective in increments of   1.0000000% defective.

USE DEFAULT SCALE? ("y" or "n")          --> y
```

Screen 2.6: Table of OC Curve (SINGLE)

```
                    OPERATING CHARACTERISTIC CURVE
                    (Type-B OC Curve for Defectives)

        Current Sampling Plan:  n =        50,   a =       1
                                tally defectives

        Percent Defective            Probability of Acceptance
        _________________            _________________________


              .0000000                     1.00000000
             1.0000000                      .91056510
             2.0000000                      .73577210
             3.0000000                      .55528060
             4.0000000                      .40048070
             5.0000000                      .27943160
             6.0000000                      .19000320
             7.0000000                      .12649350
             8.0000000                      .08271208
             9.0000000                      .05323838
            10.0000000                      .03378581

PRESS ENTER KEY TO CONTINUE
```

Table 2.3: Effect of Lot Size on the Probability of Acceptance When Using Single Sampling Plan n=50 and a=1

Type of OC Curve	Lot Size N	Percent Defective		
		1%	3%	7%
Type A	100	1.0000	0.5000	0.0559
Type A	200	0.9384	0.5324	0.0928
Type A	500	0.9194	0.5469	0.1134
Type A	1000	0.9147	0.5512	0.1200
Type A	10,000	0.9110	0.5549	0.1259
Type A	100,000	0.9106	0.5552	0.1264
Type B	-	0.9106	0.5553	0.1265

Exercises

(2.5)   Plot the OC curve of the single sampling plan n=13 and a=0 that describes this sampling plan's protection when used to inspect a lot of 100 units. Suppose defectives are tallied. What is the probability of accepting a 1% defective lot? How about an 8% or 16% defective lot?

(2.6)   In exercise 2.5, suppose the lot size changes to 10,000 units. How does this change the results?

(2.7)   Plot the OC curve of the single sampling plan n=13 and a=0 that describes this sampling plan's protection when used to inspect a series of lots. Suppose defectives are tallied. What is the probability of acceptance when the process averages 1% defective? How about 8% defective or 16% defective?

(2.8)   Plot the OC curve of the single sampling plan n=13 and a=0 that describes this sampling plan's protection when used to inspect a process. Suppose defects are tallied and the lot contains units of product. What is the probability of acceptance when the process averages 0.01 defects per unit? How about 0.08 or 0.16 defects per unit?

(2.9)   Solution tanks are inspected by pulling samples of 13 gallons and accepting if no particles are found. Plot the OC curve describing this sampling plan's protection when used to inspect a tank of 500 gallons. What is the probability of accepting a tank containing 0.01 defects per gallon? How about 0.08 or 0.16 defects per gallon?

2.4   Summarizing OC Curves

OC curves can be adequately summarized using two points on the curve called the AQL and LTPD. These two points are:

AQL   =   percent defective with a 95% chance of acceptance

LTPD[1]   =   percent defective with a 10% chance of acceptance

Figure 2.12 shows the AQL and LTPD of the single sampling plan n=50 and a=1 based on the Type-B OC curve for defectives. The AQL is 0.72%. It is determined by finding that percent defective on the bottom axis that corresponds to a probability of acceptance of 0.95 on the left axis. Likewise, the LTPD is 7.56%.

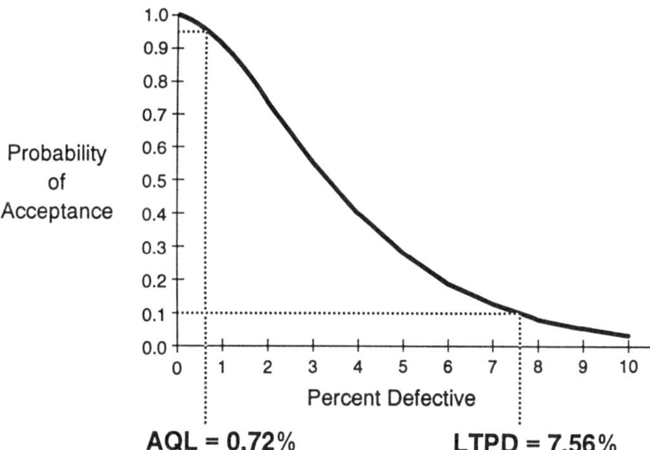

Figure 2.12: AQL and LTPD of Single Sampling Plan n=50 and a=1 Based on Type-B OC Curve for Defectives

AQL stands for acceptable quality level. It represents a level of quality that the sampling plan routinely accepts. The 5% probability of rejection at the AQL is called the producer's risk. It is the probability of falsely rejecting a good lot. The words "acceptable" and "good" are from the point of view of the sampling plan. Lots better than the AQL are routinely accepted by the sampling plan. This does

---

[1] This terminology is consistent with Schilling (1982) but is by no means standard. Some references prefer to use RQL, LQ, LQL or UQL. Certain references also prefer to use a value of 5%.

not mean they are acceptable from the customer's perspective. Sampling plans determine how to dispose of lots of product once they are produced. Once a lot has been produced, the best decision may be to release lots that are close to the AQL. A large investment in material and labor has already been made. However, from the customer's perspective, no defects are acceptable. It would have been better to have made a lot free of defects. The AQL should not be taken as permission to produce defects.

LTPD stands for lot tolerance percent defective. It represents a level of quality that the sampling plan routinely rejects. The 10% probability of acceptance at the LTPD is called the consumer's risk. It is the probability of a bad lot being falsely accepted.

Using the AQL and LTPD, the performance of the single sampling plan n=50 and a=1 can be summarized as follows:

- Lots below the AQL (0.72% defective) are routinely accepted

- Lots above the LTPD (7.56% defective) are routinely rejected

- Lots between AQL and LTPD are sometimes accepted and sometimes rejected.

The AQL and LTPD are two special cases of percentiles. The AQL is also denoted $p_{.95}$ and referred to as the 95th percentile. Likewise the LTPD is denoted $p_{.10}$ and referred to as the 10th percentile. Other percentiles can also be useful. One of these is the 50th percentile, denoted $p_{.50}$ and referred to as the indifference quality of IQ.

$$IQ \quad = \quad \text{percent defective with a 50\% chance of acceptance}$$

IQ represents a level of quality that is accepted half the time and rejected half the time. For the single sampling plan n=50 and a=1, the IQ is equal to 3.33%. Table 2.4 gives different percentiles for this sampling plan.

Table 2.4: Percentiles of the Single Sampling Plan n=50 and a=1 Based on Type-B OC Curve for Defectives

$p_{.95}$	$p_{.90}$	$p_{.50}$	$p_{.10}$	$p_{.05}$
95th percentile	90th percentile	50th percentile	10th percentile	5th percentile
AQL		IQ	LTPD, $RQL_{.10}$, $LQ_{.10}$	$RQL_{.05}$, $LQ_{.05}$
0.72%	1.07%	3.33%	7.56%	9.14%

Values of the AQL, IQ and LTPD can be obtained using option 2 of the program SINGLE. Start the program as before and use option 1 to enter the sampling plan. Then select option 2 from the main menu. Screen 2.7 will appear requesting the type of OC curve. For Type-A OC curves, the lot size is also requested. The resulting summary information is shown in Screen 2.8. The AOQL is also given. AOQLs are explained in the next section.

Screen 2.7: Input to Obtain Summary Information (SINGLE)

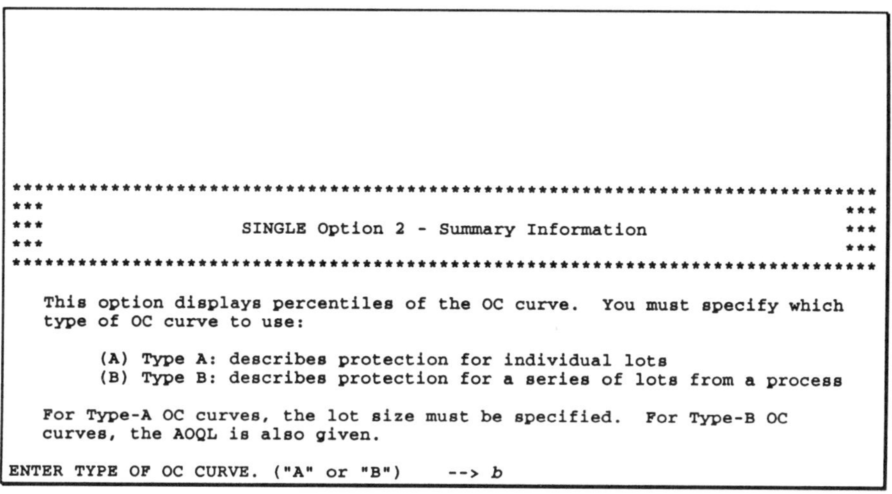

```
***********************************************************************************
***                                                                            ***
***                SINGLE Option 2 - Summary Information                       ***
***                                                                            ***
***********************************************************************************

    This option displays percentiles of the OC curve.  You must specify which
    type of OC curve to use:

        (A) Type A: describes protection for individual lots
        (B) Type B: describes protection for a series of lots from a process

    For Type-A OC curves, the lot size must be specified.  For Type-B OC
    curves, the AOQL is also given.

ENTER TYPE OF OC CURVE. ("A" or "B")     --> b
```

Screen 2.8: Summary Information (SINGLE)

```
                        SUMMARY STATISTICS
                   (Type-B OC Curve for Defectives)

        Current Sampling Plan:  n =          50,  a =       1
                                tally defectives


                                              Process
                Percentile     Symbol     Percent Defective
                __________     ______     _________________

                  95.0%         AQL            .7153718
                  90.0%                        1.0686790
                  50.0%         IQ             3.3340260
                  10.0%         LTPD           7.5580570
                   5.0%                        9.1398150


        AOQL =      1.6697490%     (Assumes 100% inspection of rejected
                                    lots with all defectives being found
                                    and repaired plus a large lot size.)

PRESS ENTER KEY TO CONTINUE
```

For Type-B OC curves, the OC curve is smooth. Therefore the AQL and LTPD are well defined. For Type-A OC curves, the OC curve is a series of points. No lot quality may have exactly a 95% chance of acceptance. To obtain AQLs and LTPDs for Type-A OC curves, the points on the OC curve are first connected by straight line segments. Figure 2.13 shows how.

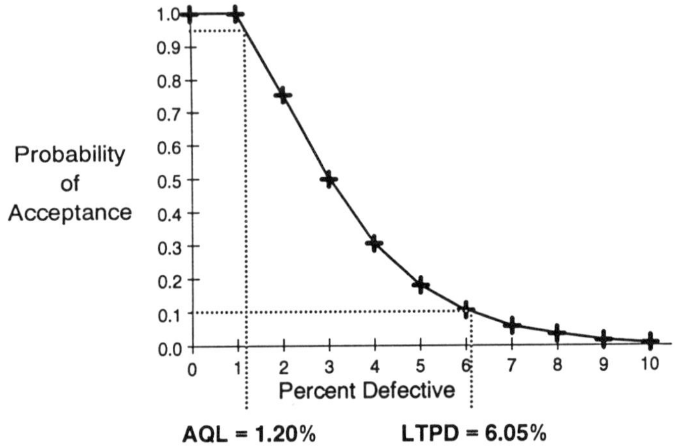

Figure 2.13: AQL and LTPD of Single Sampling Plan n=50 and a=1 Based on Type-A OC Curve for Defectives with N=100

For Type-A OC curves, the probability of acceptance depends, at least to some degree, on the lot size. As a result, the AQL and LTPD also depend on the lot size. Table 2.5 shows the effect of lot size on the AQL and LTPD for the single sampling plan n=50 and a=1 when defectives are tallied. For large lot sizes, the values rapidly approach the AQL and LTPD of the Type-B OC curve.

Once the lot size is at least 10 times the sample size, the AQL and LTPD change little. As a result:

> **When the sample size is small compared to the lot size, the protection afforded by acceptance sampling is not effected by the lot size. It is solely the result of the sample size and accept number.**

This fact surprises many. They mistakenly believe that it is the percentage of the lot sampled that determines the protection. It is not. The protection is almost exclusively a result of the sampling plan used. The single sampling plan n=50 and a=1 offers the same protection for a lot of 100,000 units (0.05% inspected) as it does for a lot of 500 units (10% inspected). This is true whether one is tallying defects or defectives.

Table 2.5: Effect of Lot Size on the AQL and LTPD of the Single Sampling Plan n=50 and a=1 When Defectives are Tallied

Type of OC Curve	Lot Size N	AQL	LTPD
Type A	100	1.202	6.048
Type A	200	0.906	6.858
Type A	500	0.787	7.284
Type A	1000	0.749	7.422
Type A	10,000	0.719	7.545
Type A	100,000	0.716	7.557
Type B	-	0.715	7.558

Exercises

(2.10) What are the AQL and LTPD of the single sampling plan n=13 and a=0 when used to inspect a lot of 100 units? Assume defectives are tallied.

(2.11) What are the AQL and LTPD of the single sampling plan n=13 and a=0 when used to inspect a lot of 10,000 units? Assume defectives are tallied.

(2.12) What are the AQL and LTPD of the single sampling plan n=13 and a=0 when used to inspect a series of lots from a process? Assume defectives are tallied.

(2.13) What are the AQL and LTPD of the single sampling plan n=13 and a=0 when used to inspect a lot of 100 units? Assume defects are tallied.

(2.14) What are the AQL and LTPD of the single sampling plan n=13 and a=0 when used to inspect a series of solution tanks? Assume defects are tallied.

2.5   AOQ Curves and AOQL

Customers are concerned with the quality of the goods they receive. They want protection against highly defective lots. The LTPD describes a sampling plan's protection against such lots. However, if the customer receives a series of lots; of greater importance is the average level of defects received over the long haul. This is referred to as the average outgoing quality or AOQ for short. The average outgoing quality is the average quality of all released lots. It is the quality of the product released from the inspection station. The average outgoing quality depends on the quality of the product coming into the inspection station, the particular sampling plan used, and the method of handling rejected lots.

Sampling plans by themselves cannot provide any guarantee as to the average outgoing quality. Suppose a process produces a long series of 5% defective lots. If rejected lots are discarded, then, even though the sampling plan may reject many of the lots, all released lots are still 5% defective. In this case, the average outgoing quality is the same as the average incoming quality, namely 5%. When rejected lots are discarded, acceptance sampling may do little or nothing to improve the outgoing quality.

By combining acceptance sampling with 100% inspection of rejected lots, assurances can be made as to the average outgoing quality. To see how, look at the AOQ curve given in Figure 2.14. This curve is for the single sampling plan n=50 and a=1 when defectives are tallied. The bottom axis is the process percent defective, i.e., the incoming quality. Given the incoming quality on the bottom axis, the AOQ curve gives the resulting outgoing quality. If the incoming quality is 1% defective, the average outgoing quality is 0.91% defective. Likewise, when the incoming quality is 5% defective, the AOQ is 1.40% defective. The AOQ curve starts low as few defectives are made. It ends low as most lots are 100% inspected. It reaches a maximum somewhere in between.

Of particular interest is the maximum value of the AOQ curve. This maximum value is called the average outgoing quality limit or AOQL. The AOQL for the single sampling plan n=50 and a=1 is 1.67% defective. Regardless of how the incoming quality behaves, the sampling plan plus 100% inspection of rejected lots ensures the average quality going to the customer is no worse than the AOQL. The AOQL of a sampling plan can be determined by using option 2 of the program SINGLE (See Screen 2.8 on page 25).

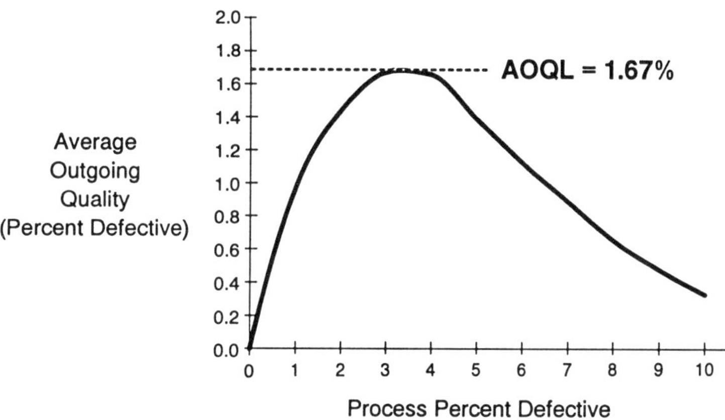

**Figure 2.14:   AOQ Curve of Single Sampling Plan n=50 and a=1
When the Number of Defectives is Tallied**

The AOQ curve and AOQL in Figure 2.14 where calculated under the following assumptions:

- Rejected lots are 100% inspected

- The 100% inspection finds all defective units

- Defectives found by the inspection are repaired or replaced

- The lot size is large compared to the sample size

Under these assumptions, the average outgoing quality is:

$$\text{AOQ(p)} \;=\; p \times P_a(p)$$

where:

$$p \qquad = \qquad \text{incoming percent defective}$$

$$P_a(p) \;=\; \text{probability of accepting lots from a process that is}$$
$$\text{p percent defective}$$

To see the basis of this formula, assume one is inspecting a series of 5% defective lots using the single sampling plan n=50 and a=1. At 5% defective, the sampling plan accepts 27.9% of the lots. These lots will be released as they are. The other 72.1% of the lots will be rejected and 100% inspected. When released, they are 0% defective. The outgoing quality is therefore:

$$\text{AOQ(5\%)} \;=\; (5\%) \times 0.279 + (0\%) \times 0.721 \;=\; 1.40\% \text{ defective.}$$

$$=\; (5\%) \times 0.279 \;=\; p \times P_a(p)$$

This is a sizable improvement in the outgoing quality at the cost of frequent 100% inspections. When defects are tallied, replace the p in the above formula with an r where r is the average number of defects per unit for the process.

Option 2 of the program SINGLE uses the assumptions given above to calculate the AOQL. These assumptions are very restrictive. Further, violation of these assumptions can have a large effect on the AOQ curve. The user is reminded of these assumptions with the message in Screen 2.8 following the AOQL value. AOQLs given elsewhere including Mil-Std-105E make these same assumptions.

Option 4 of program SINGLE can be used to plot or tabulate AOQ curves. Option 4 can generate curves that make none of the previous assumptions except that rejected lots are 100% inspected. This requires that one input the lot size, the efficiency of the 100% inspection, and specify whether defectives are repaired, replaced or discarded.

When defectives are tallied, option 4 uses the following more complex formulas for the AOQ(p):

$$\text{AOQ}_{\text{repair or replace unit}}(p) \;=\; p \times \left(1 - \frac{n}{N}\right) \times \left[1 - \frac{E}{100} \times \left(1 - P_a(p)\right)\right]$$

$$\text{AOQ}_{\text{discard unit}}(p) \;=\; \frac{\text{AOQ}_{\text{repair or replace unit}}(p)}{1 - \dfrac{n}{N}\dfrac{p}{100} - \left(1 - \dfrac{n}{N}\right) \times \dfrac{E}{100} \times \dfrac{p}{100} \times \left(1 - P_a(p)\right)}$$

where n is the sample size, N is the lot size, and E is the efficiency of 100% inspection as a percentage. The first formula gives the AOQ when defective units are repaired or replaced. The second formula covers the case when defective units are discarded. Further information can be found in Beainy and Case (1981).

Screen 2.9 shows the input of the information required by option 4. The single sampling plan n=50 and a=1 where defectives are tallied has already been entered. A value of 100% was entered for the efficiency of the 100% inspection. Zero was entered for the lot size indicating the lot size is large. This results in n/N being set to zero in the above formulas. Finally, replacement of defectives was specified. These are exactly the conditions assumed for calculating the AOQL under option 2. A plot of the AOQ curve was requested using the default scale. This plot is shown in Screen 2.10. Notice how the AOQ reaches a maximum close to 1.7%. This is the AOQL. From Screen 2.8, the exact value of the AOQL is 1.6697490%.

Screen 2.9: Entry of AOQ Curve Information (SINGLE)

```
*******************************************************************************
***                                                                         ***
***                    SINGLE Option 4 - AOQ Curve                          ***
***                                                                         ***
*******************************************************************************

    This option plots or tabulates the AOQ curve.  It is assumed that
    rejected lots are 100% inspected.  One must enter the efficiency of
    the 100% inspection, the lot size, and whether defects and
    defectives found are either repaired, replaced or discarded.
    One can specify the scale or let the program select the appropriate scale.

ENTER EFFICIENCY OF 100% INSPECTION      --> 100
ENTER LOT SIZE (0 = Large)               --> 0
DEFECTS? (1=repair,2=replace,3=discard)  --> 2
ENTER "p" FOR PLOT "t" FOR TABLE.        --> p

    The default scale for percent defective (bottom axis) is from 0%
    to  10.0000000% defective.

USE DEFAULT SCALE? ("y" or "n")          --> y

    The default scale for average outgoing quality (left axis) is from 0%
    to   2.5000000% defective.

USE DEFAULT SCALE? ("y" or "n")          --> y
```

Screen 2.10: Plot of AOQ Curve (SINGLE)

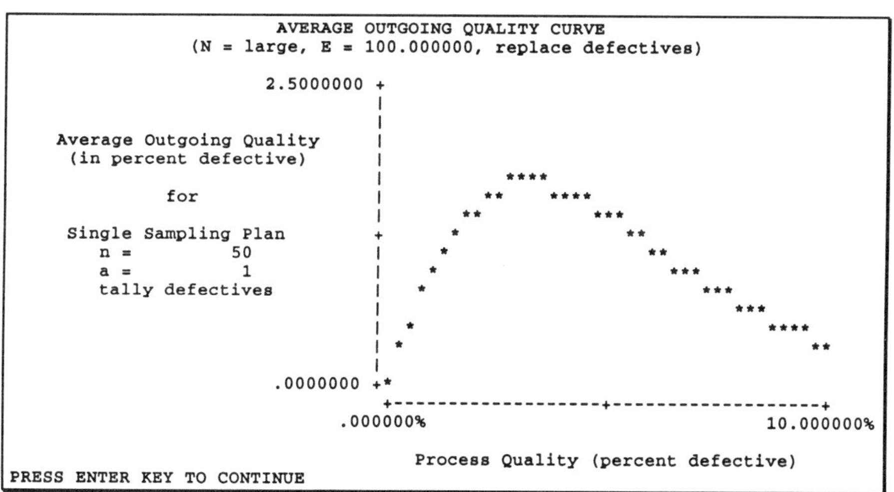

A table of the AOQ curve can also be obtained using option 4. Screen
2.11 shows the corresponding table for the AOQ curve shown in
Screen 2.10. This table uses the default scale.

Screen 2.11: Table of AOQ Curve (SINGLE)

```
                    AVERAGE OUTGOING QUALITY CURVE
          (N = large, E = 100.000000, replace defectives)

    Current Sampling Plan:   n =          50,   a =       1
                             tally defectives


        Percent Defective            Average Outgoing Quality
    ________________________         ________________________

              .0000000                        .0000000
             1.0000000                        .9105651
             2.0000000                       1.4715440
             3.0000000                       1.6658420
             4.0000000                       1.6019230
             5.0000000                       1.3971580
             6.0000000                       1.1400190
             7.0000000                        .8854554
             8.0000000                        .6616960
             9.0000000                        .4791455
            10.0000000                        .3378588

    PRESS ENTER KEY TO CONTINUE
```

Figure 2.15 shows how changing the efficiency of the 100% inspection changes the shape of the AOQ curve. The AOQ curve is very sensitive to the efficiency of the 100% inspection. Therefore, the AOQL is only relevant when rejected lots are 100% inspected and this 100% inspection is extremely efficient, i.e., 95% or better.

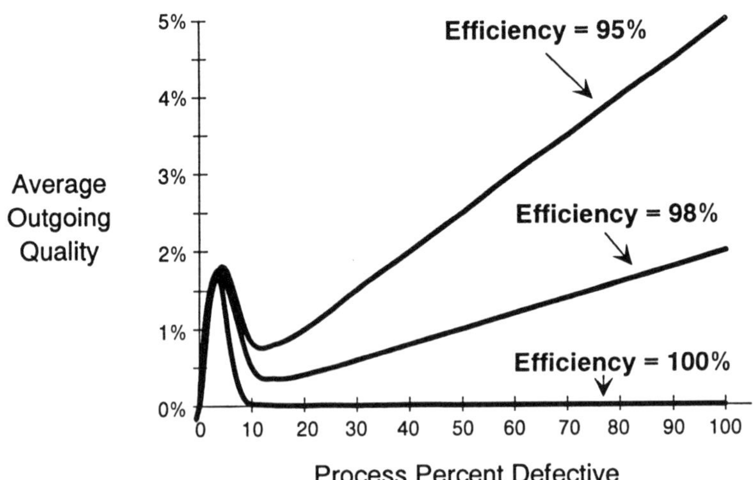

Figure 2.15: AOQ Curves for Different Efficiencies
Single Sampling Plan n=50 and a=1

When defects are tallied, the formulas for AOQ are:

$$AOQ_{\text{repair defect}}(r) \;=\; r \times \left(1 - \frac{n}{N}\right) \times \left[1 - \frac{E}{100} \times (1 - P_a(r))\right]$$

$$AOQ_{\text{replace unit}}(r) \;=\; r \times \left(1 - \frac{n}{N}\right) \times \left[1 - \left(\frac{1 - e^{-r\frac{E}{100}}}{1 - e^{-r}}\right) \times (1 - P_a(r))\right]$$

$$AOQ_{\text{discard unit}}(r) \;=\; \frac{AOQ_{\text{replace unit}}(r)}{1 - \frac{n}{N}\left(1 - e^{-r}\right) - \left(1 - \frac{n}{N}\right) \times \left(1 - e^{-r\frac{E}{100}}\right) \times (1 - P_a(p))}$$

where:

r = average number of defects per unit in incoming lots

$P_a(r)$ = probability of accepting lots from a process averaging r defects per unit.

n = sample size

N = lot size

E = efficiency of 100% inspection as a percentage

The first formula gives the AOQ when defects found are repaired. The second formula gives the AQL when units containing defects are replaced by units containing no defects. To see the difference, assume one is inspecting a section of an airplane wing for defective rivets. In the first case, finding a defective rivet only results in the repair of that rivet. In the second case, the whole wing section is replaced resulting in the elimination of all defects on it.. The second case removes more defects since a defect that was initially missed may be replaced anyway when the unit it is on is replaced as a result of some other defect. The third formula covers the case when units containing one or more defects are discarded. If the lot is continuous (tank, load of cement, etc.), then the first formula is always used. In this case, defects identified are simply removed.

Exercise

(2.15) Obtain a plot the AOQ curve for the single sampling plan n=13, a=0 when defects are tallied, the lot consists of individual units, the lot size is large, the 100% inspection is 99% efficient, and defective units are discarded. What is the AOQ at 0.005, 0.01 and 0.05 defects per unit?

2.6   ASN Curves

Single sampling plans select n units from each lot. The number of units inspected does not vary. However, sometimes the accept/reject decision is determined before all n units are inspected. Suppose the single sampling plan n=50 and a=1 is used. If the first ten units inspected contain 2 defective units, then the lot will be rejected regardless of the results of the remaining 40 units. Alternatively, if none of the first 49 units are defective, then the lot will be accepted regardless of the result of the last unit. To reduce the amount of inspection, one can curtail inspection once the accept/reject decision is determined.

Inspection can be curtailed and the lot rejected as soon as a+1 defects or defective units are found. Inspection can be curtailed and the lot accepted as soon as n-a nondefective units are found. When defects are tallied, inspection can not be curtailed on acceptance. This is because more than one defect can occur per unit.

When inspection is curtailed, the actual number of units inspected will vary from lot to lot. Of interest is the average number of units inspected. The average number of units inspected depends on the process quality. A plot of the average number of units inspected versus process quality is shown in Figure 2.16. Such a plot is call an average sample number curve or ASN curve.

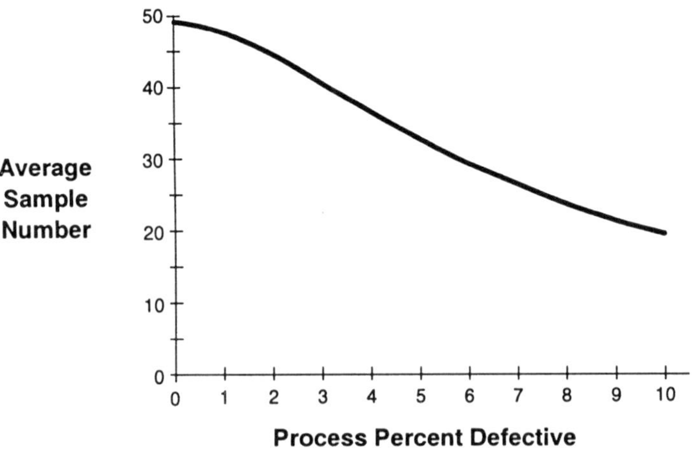

**Figure 2.16: ASN Curve of Single Sampling Plan n=50 and a=1
Defectives Tallied, Inspection Fully Curtailed**

Curtailing of inspection does not effect the protection provided by a sampling plan. The same lots are accepted and rejected as before. The OC curve, AOQ curve, AQL, IQ, LTPD and AOQL are all unchanged.

The program SINGLE can be used to obtain a plot or table of the ASN curve. An ASN curve is obtained by selecting option 5 from the main menu. Selecting option 5 results in Screen 2.12. This screen requests the method of curtailing, whether a plot or table is desired, and information on scaling. Screen 2.12 shows the input required to obtain a plot of the ASN curve for full curtailing (both on acceptance and rejection) using the default scale. The resulting plot is shown in Screen 2.13. A table of the same OC curve is given in Screen 2.14. The formulas used to calculate the ASN are given in Appendix C.

Screen 2.12: Entering ASN Curve Information (SINGLE)

```
***********************************************************************************
***                                                                          ***
***                   SINGLE Option 5 - ASN Curve                            ***
***                                                                          ***
***********************************************************************************

   This option plots or tabulates the ASN curve.  The method of curtailing must
   be specified.  Curtailing on rejection stops when a+1 defects or defectives
   are found.  Curtailing on acceptance stops when n-a non defectives are
   found.  Curtailing on acceptance is not possible when defects are tallied.
   One may specify the scale or let the program select the appropriate scale.

CURTAILING ON REJECTION?   ("y" or "n")   --> y
CURTAILING ON ACCEPTANCE?  ("y" or "n")   --> y
ENTER "p" FOR PLOT "t" FOR TABLE.         --> p

   The default scale for percent defective (bottom axis) is from 0%
   to  10.0000000% defective.

USE DEFAULT SCALE? ("y" or "n")           --> y

   The default scale for average sample number (left axis) is from 0
   to        50.00000.

USE DEFAULT SCALE? ("y" or "n")           --> y
```

Both AOQ and ASN curves have process quality on the bottom axis, the same as Type-B OC curves. The AOQ and ASN are long term averages. They only apply to the inspection of a series of lots.

While curtailing inspection reduces inspection costs, it increases the complexity. The savings are generally minimal, especially when most lots are of good quality. Curtailing can also interfere with other uses of the inspection data such as control charts and tracking the process quality. For these reasons, single sampling plans are rarely curtailed.

Screen 2.13: Plot of ASN Curve (SINGLE)

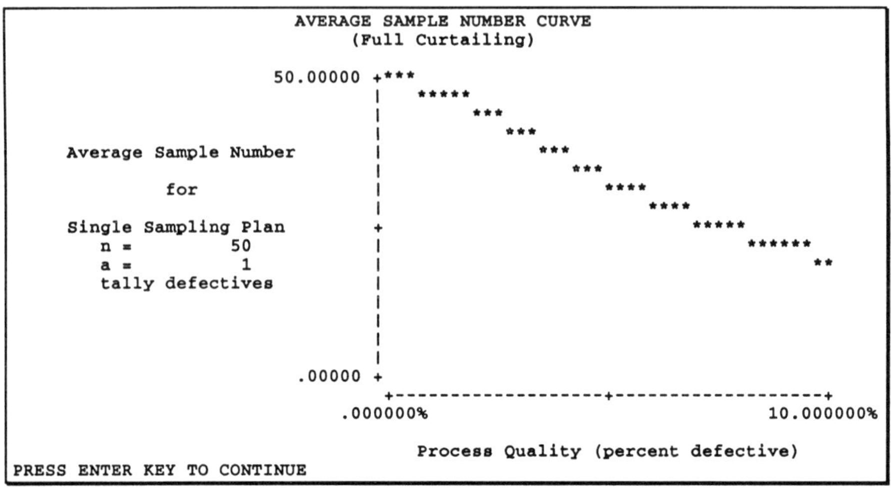

Screen 2.14: Table of ASN Curve (SINGLE)

```
                    AVERAGE SAMPLE NUMBER CURVE
                         (Full Curtailing)

     Current Sampling Plan:  n =         50,   a =      1
                             tally defectives

         Percent Defective           Average Sample Number
        ___________________         _______________________

                 .0000000                       49.00000
                1.0000000                       47.83174
                2.0000000                       44.63129
                3.0000000                       40.66365
                4.0000000                       36.60555
                5.0000000                       32.79148
                6.0000000                       29.36287
                7.0000000                       26.35646
                8.0000000                       23.75596
                9.0000000                       21.52134
               10.0000000                       19.60488

PRESS ENTER KEY TO CONTINUE
```

Up to now we have seen how to evaluate single sampling plans.
Options 2, 3, 4 and 5 can be used to obtain the OC, AOQ and ASN
curves as well as the AQL, IQ, LTPD and AOQL. To those not
familiar with acceptance sampling, understanding the protection
provided by one's sampling plans can be a real eye opener. Often the
protection provided is less than believed. Many times the desired
protection cannot be obtained at a reasonable cost.

Some hold the false belief that no matter what else happens, the sampling plan will make it all right. For these individuals, understanding the actual protection provided by acceptance sampling can lead to many questions concerning the quality going to the customer and the benefits of acceptance sampling. It can lead to the ultimate realization that the best way of ensuring that the customer gets good quality is to make it. Sometimes, the best way to get improvement efforts started is to remove the acceptance sampling crutch. Next we turn our attention to selecting single sampling plans.

Exercises

(2.16)   Plot the ASN curve of the single sampling plan n=50 and a=1 when defectives are tallied and inspection is curtailed on rejection only. What is the ASN at 1% and 5% defective?

(2.17)   Plot the ASN curve of the single sampling plan n=50 and a=1 when defectives are tallied and inspection is curtailed on acceptance only. What is the ASN at 1% and 5% defective?

(2.18)   Plot the ASN curve of the single sampling plan n=50 and a=1 when defects are tallied, the lot is continuous, and inspection is curtailed on rejection. What is the ASN at 0.01 and 0.05 defects per unit?

2.7   Summary

Single sampling plans have two parameters: the sample size n and the accept number a. Acceptance is based on a tally of either the number of defective units or the number of defects. Lots whose tally is less than or equal to the accept number are accepted.

OC curves are plots of the probability of acceptance versus the process or lot quality. They summarize the protection provided by sampling plans. An OC curve shows what quality lots are routinely accepted and rejected. Ideally, all good quality lots should be accepted and all bad quality lots rejected. In practice a compromise must be made between the cost of performing the inspection and the ability to distinguish between good and bad quality lots.

There are Type-A and Type-B OC curves for defectives and defects. Type-A OC curves describe the protection provided by sampling plans for individual lots. Type-B OC curves describe the protection for a series of lots from a process. Type-A OC curves converge to Type-B OC curves as the lot size increases.

OC curves can be summarized using the AQL, LTPD and IQ. Lots better than the AQL are consistently accepted. Lots worse than the LTPD are consistently rejected. Lots between the AQL and LTPD are sometimes accepted and sometimes rejected. Lots at the IQ are accepted half of the time. The AQL, LTPD and IQ are special cases of percentiles.

So long as the lot size is at least 10 times the sample size, the lot size has little effect on the AQL and LTPD. The protection provided by a sampling plan is primarily determined by the number of samples selected and the accept number. The lot size or percentage of the lot inspected has little effect. Fifty samples provides the same protection when inspecting a lot of 500 units (10%) as it does when applied to a lot of 10,000 units (0.5%).

When rejected lots are 100% inspected, the AOQ curve and AOQL can be determined. The commonly used formulas for AOQ and AOQL assume that rejected lots are 100% inspected, the 100% inspection is 100% effective, the lot size is large compared to the sample size, and defective units found as a result of the sampling or 100% inspection are reworked or replaced. Of these assumptions, the assumption that the 100% inspection is 100% effective is crucial. AOQL values should not be used unless the 100% inspection has an efficiency close to 100%. AOQ curves under varying assumptions can be generated to examine the effect of these assumptions.

Curtailing of inspection can be performed in order to reduce the average number of units inspected. Inspection can be curtailed on rejection as soon as a+1 defective units or defects are observed. It can be curtailed on acceptance as soon as n-a nondefective units are observed. An ASN curve is a plot of the average number of units inspected versus process quality. Curtailing of inspection may conflict with other uses of the inspection results such as control charting and tracking the process average. As a result, it is not commonly performed.

References

Beainy, Ilham and Case, Kenneth E. (1981). "A Wide Variety of AOQ and ATI Performance Measures With and Without Inspection Error." Journal of Quality Technologies, Volume 13, Pages 1-9.

Schilling, Edward G. (1982). Acceptance Sampling in Quality Control. Marcel Dekker, New York, New York.

3

Selecting
Single Sampling Plans

The objective of this chapter is to:

- Learn how to select single sampling plans based on the protection they provide.

Sampling plans can also be selected based on economics. The selection of economic based sampling plans is covered in Chapter 4.

Single sampling plans are but one type of sampling plan. Other types include double sampling plans, quick switching systems, and variables sampling plans. These are covered in Chapters 6, 7 and 9 respectively. These alternative types of sampling plans can frequently provide the same protection as single sampling plans while reducing the number of units inspected. However, they also increase the administrative complexity and may not always be applicable.

Selecting the best sampling plan for the job requires selecting the best single sampling plan, the best double sampling plan, the best quick switching system, and the best variables sampling plan. These sampling plans are then compared based on the number of units inspected, administrative complexity, and applicability to determine the best overall plan. This chapter starts this process by providing procedures for selecting the best single sampling plan.

3.1   Selecting Plans Based on AQL and LTPD

One use of acceptance sampling is as an insurance policy against the release of highly defective lots. For this application, most lots produced are acceptable. However, there is concern over the occasional release of a highly defective lot. Acceptance sampling can provide protection against the release of such highly defective lots. It is akin to taking out an insurance policy. The premiums are the samples taken from each lot. The payoff is the rejection of a highly defective lot. The premiums are low but must be routinely paid. The payoffs are infrequent but avoid potentially large losses.

When using acceptance sampling as an insurance policy against highly defective lots, one needs to specify the desired level of protection. This can be done by specifying the desired LTPD. The sampling plan then ensures that lots with defect rates above the LTPD are routinely rejected. Following acceptance of a lot, one can state that with 90% confidence, the lot's percent defective or defect rate is below the LTPD.

One also wants to ensure that good lots are released. This is accomplished by selecting a sampling plan whose AQL is above the historical average for lot quality. Setting the AQL equal to the historical average would result in a rather high reject rate of 5%. Keeping the premiums low requires minimizing the number of samples inspected. This is accomplished by selecting the sampling plan with the specified AQL and LTPD that minimizes the sample size (See Figure 3.1). When guarding against the release of highly defective lots, one wants protection on a lot by lot basis. Therefore, Type-A OC curves should be used.

(1)   Specify LTPD representing the level of defects or defectives to guard against.

(2)   Specify AQL above the historical average for lot quality.

(3)   Select sampling plan minimizing the sample size where:

actual AQL $\geq$ specified AQL

actual LTPD $\leq$ specified LTPD

**Figure 3.1: Procedure for Selecting Sampling Plans
Guarding Against Highly Defective Lots**

Once one has selected an AQL and LTPD, the program SINGLE can be used to select the appropriate single sampling plan. This single sampling plan should have the smallest sample size of those single sampling plans satisfying the above conditions. Begin by starting the program and selecting option 6 from the main menu. Screen 3.1 will appear requesting further information including the AQL and LTPD.

Suppose one wishes to select a single sampling plan with an AQL of 1% and LTPD of 10% where defectives are tallied. Screen 3.1 shows the required input. A Type-A OC curve is requested requiring that the lot size be input. A lot size of 1000 is specified. The last question is whether to use the standard definitions of the AQL and LTPD. The standard definition of the AQL is that it has a 95% chance of acceptance. The standard definition of the LTPD is that it has a 10% chance of acceptance. Responding no to this question allows one to specify other values. A value of 5% for the LTPD is not uncommon.

Screen 3.1: Selection of Single Sampling Plans
Based on AQL and LTPD (SINGLE)

```
****************************************************************************
***                                                                      ***
***       SINGLE Option 6 - Select Sampling Plan Based on AQL and LTPD    ***
***                                                                      ***
****************************************************************************

   This option selects the single sampling plan minimizing the ASN curve
   from among those satisfying the following set of conditions:

      Actual AQL is greater than or equal to the specified AQL
      Actual LTPD is less than or equal to the specified LTPD

   One must specify whether to use the Type-A or Type-B OC curve for
   defects or defectives.  The standard definitions of AQL (95% chance of
   acceptance) and LTPD (10% chance of acceptance) are used unless indicated
   otherwise.

TALLY? (1=defectives,2=defects)          --> 1
ENTER TYPE OF OC CURVE. ("A" or "B")     --> a
ENTER LOT SIZE                           --> 1000
ENTER AQL  AS PERCENT DEFECTIVE          --> 1
ENTER LTPD AS PERCENT DEFECTIVE          --> 10
USE .95 & .1 FOR AQL-LTPD? ("y" or "n") --> y
```

When all the questions have been answered, pressing the enter key results in the sampling plan being selected and displayed as in Screen 3.2. The single sampling plan selected is n=37 and a=1. The algorithms used by the program SINGLE are detailed in Guenther (1977) and Taylor (1983).

Screen 3.2: Results of Selecting the Single Sampling Plan with AQL=1%, LTPD=10%, Tally Defectives Using Type-A OC Curve With N=1000   (SINGLE)

```
********************************************************************************
*                            *                                                *
*                            *        Copyright (C) 1992 Taylor Enterprises   *
*     Program SINGLE         *         P.O. Box 820, Lake Villa, IL 60046     *
*                            *                 (708) 356-1074                 *
*                            *                                                *
********************************************************************************

              Current Sampling Plan:  n =          37,   a =     1
                                       tally defectives


                          -----  MENU OPTIONS  -----

                  (1) Enter sampling plan
                  (2) Summary information
                  (3) OC curve
                  (4) AOQ curve
                  (5) ASN curve
                  (6) Select sampling plan based on AQL and LTPD
                  (7) Select sampling plan based on AQL and AOQL
                  (8) Select sampling plan based on AQL and n

ENTER NUMBER OF OPTION OR "q" TO QUIT  -->
```

Option 2 can be used to examine the selected sampling plan. This plan has an actual AQL of 1.003% and an actual LTPD of 9.978%. These satisfy the required conditions:

Actual AQL (1.003%) $\geq$ Specified AQL (1%)

Actual LTPD (9.978%) $\leq$ Specified LTPD (10%)

These conditions ensure that the actual OC curve be at least as discriminating as the specified OC curve. Of the numerous single sampling plans satisfying the above conditions, the one minimizing the sample size was selected. This sampling plan also minimizes n-a and a+1. These are the stopping rules for curtailed inspection. As a result, this single sampling plan minimizes the ASN over the entire ASN curve regardless of the method of curtailing.

When defects are tallied, the AQL and LTPD are specified as defect rates rather than percent defectives. When inspecting from a continuous lot, one must also specify the smallest sample size increment. For example, specifying 0.1 means allowable sample sizes are 0.1, 0.2, 0.3, ···. Alternatively, specifying 5.0 means allowable sample sizes are 5.0, 10.0, 15.0, ···. Suppose one is inspecting loads of cement for large stones and foreign objects. An AQL of 1 defect per cubic yard and an LTPD of 20 defects per cubic yard is desired. A nine cubic inch sampler is used to select samples. Translating this to cubic yards results in a smallest sample size increment of 0.006944444 cubic yards. Since the sampling plan must provide the desired protection for each load of cement inspected, a Type-A OC curve should be used. The lot size is 5 cubic yards. Screen 3.3 shows the required input. The selected sampling plan has n = 0.19444 cubic yards and a=1. This corresponds to 28 samples from the 9 cubic inch sampler.

Screen 3.3: Screen for Selecting Plan When Defects are Tallied and the Lot is Continuous (SINGLE)

```
*************************************************************************************
***                                                                             ***
***        SINGLE Option 6 - Select Sampling Plan Based on AQL and LTPD         ***
***                                                                             ***
*************************************************************************************

    This option selects the single sampling plan minimizing the ASN curve
    from among those satisfying the following set of conditions:

        Actual AQL is greater than or equal to the specified AQL
        Actual LTPD is less than or equal to the specified LTPD

    One must specify whether to use the Type-A or Type-B OC curve for
    defects or defectives.  The standard definitions of AQL (95% chance of
    acceptance) and LTPD (10% chance of acceptance) are used unless indicated
    otherwise.

TALLY? (1=defectives,2=defects)          --> 2
LOT? (1=units, 2=continuous)             --> 2

    Non integer sample sizes are possible.  You must specify the smallest
    possible increment for the sample size, say x.  Then the possible samples
    sizes are x, 2x, 3x, ... .

ENTER SMALLEST SAMPLE SIZE INCREMENT     --> 0.00694444
ENTER TYPE OF OC CURVE. ("A" or "B")     --> a
ENTER LOT SIZE                           --> 5
ENTER AQL  AS DEFECT RATE                --> 1
ENTER LTPD AS DEFECT RATE                --> 20
USE .95 & .1 FOR AQL-LTPD? ("y" or "n") --> y
```

The sampling plans selected have all been based on Type-A OC curves. This is because lot by lot protection is desired. However, it makes the sampling plan dependent on lot size. What should one do if the lot size is unknown or varies? Table 3.1 shows the effect of the lot size on the sampling plan selected. The last sampling plan in Table 3.1 was selected using the Type-B OC curve. As the lot size increases, the sampling plans converge to the sampling plan selected using the Type B-OC curve. When the lot size is large, single sampling plans can be selected based on the Type-B OC curve. In which case, the lot size does not have to be specified.

**Table 3.1: Effect of Lot Size on the Sampling Plan Selected
For an AQL = 1.0% and LTPD = 10.0%
When Defectives are Tallied**

OC Curve	Sampling Plan	
	n	a
Type A, N=100	33	1
Type A, N=200	35	1
Type A, N=500	37	1
Type A, N=1,000	37	1
Type A, N=10,000	52	2
Type A, N=100,000	52	2
Type B	52	2

**Table 3.2: Probability of Acceptance at Specified AQL and LTPD
Single Sampling Plan n=52 and a=2, Defectives Tallied**

OC Curve	Probability of Acceptance	
	At AQL = 1%	At LTPD = 10%
Type A, N=100	1.0000	0.0343
Type A, N=200	1.0000	0.0660
Type A, N=500	0.9908	0.0847
Type A, N=1,000	0.9877	0.0907
Type A, N=10,000	0.9849	0.0960
Type A, N=100,000	0.9847	0.0966
Type B	0.9846	0.0966

What happens when a sampling plan selected based on its Type-B OC curve is used for smaller lot sizes? Table 3.2 shows the effect of lot size on the probability of acceptance for the single sampling plan n=52 and a=1. This sampling plan was selected by specifying an AQL of 1% and LTPD of 10% based on the Type-B OC curve. At the specified AQL of 1%, the probabilities of acceptance for the Type A OC curves are all greater than the probability of acceptance for the Type-B OC curve. In all cases the probability of acceptance is at least 0.95. Therefore, in all cases, the actual AQL is above 1%.

At the specified LTPD of 10%, just the opposite is true. The Type-A OC curves have probabilities of acceptance less than that of the Type-B OC curve. In all cases, the probability of acceptance is less than 0.1. Therefore, in all cases, the actual LTPD is below 10%.

The single sampling plan n=52 and a=1 was selected to satisfy the following conditions based on the Type-B OC curve:

$$\text{Actual AQL} \geq \text{Specified AQL}$$

$$\text{Actual LTPD} \leq \text{Specified LTPD}$$

It also satisfies these same conditions for Type-A OC curves, regardless of lot size. This is a result of the fact that Type-A OC curves are steeper than Type-B OC curves (See Figure 3.2). One can select sampling plans based on Type-B OC curves and then apply them to individual lots without lessening the protection.

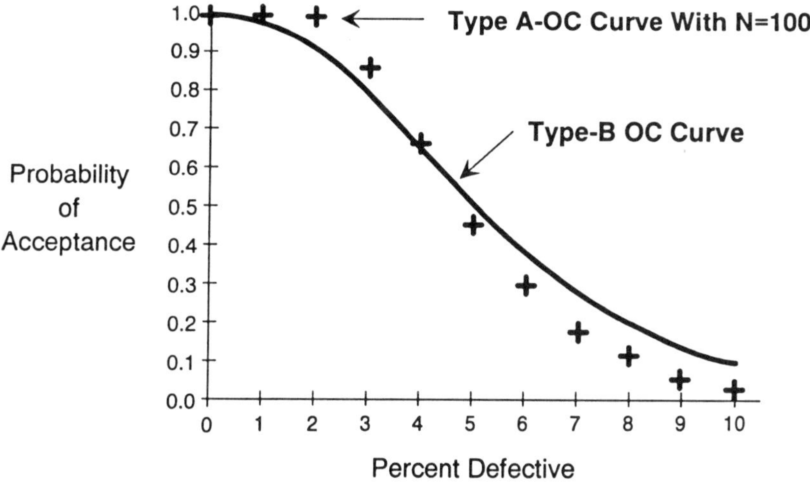

Figure 3.2: OC Curves for Single Sampling Plan n=52 and a=2

An alternate approach to selecting single sampling plans with a specified AQL and LTPD is to use tables. Most tables, including the ones in this book, are based on Type-B OC curves. This ensures that the sampling plan selected provides the desired protection regardless of the lot size. The downside to this approach is that the resulting sample size may be larger than required for smaller lot sizes. In Table 3.1, if one knew that the lot size would always be below 1000, the single sampling plan n=37 and a=1 could be used instead of n=52 and a=2.

Table 3.3 gives over a hundred single sampling plans indexed by their AQL and LTPD. This table covers sample sizes up to 2000 units and accept numbers up to 18 where the AQL is less than or equal to 10%. It is assumed that defectives are tallied.

To use Table 3.3, select the desired AQL from the first column. Then scan across the row to find the sampling plan with the desired LTPD. For example, to select a plan with an AQL of 1.0% and LTPD of 10%, find the row labeled with an AQL of 1.0%. The second plan in this row, "32/(1,2)", is the one that comes closest to the desired LTPD. It has an LTPD of 11.6%. The expression "32/(1,2)" is compact notation for the single sampling plan n=32 and a=1. The number two is the reject number which is always one greater than the accept number.

When the program SINGLE was used to select a single sampling plan with an AQL of 1.0% and LTPD of 10%, the resulting plan was n=52 and a=2. Table 3.3 gave us a different plan where n=32 and a=1. Why the difference? The first plan, n=52, satisfies the conditions AQL ≥ 1.0% and LTPD ≤ 10%. The second plan does not strictly satisfy these conditions, but comes close. Missing by a small amount can result is a dramatic increase in the sample size.

In Table 3.3, all sampling plans in the same column have the same accept number. Single sampling plans with the same accept number will have similar ratios of the LTPD to the AQL. These ratios are given at the top of each column. Since the accept number is limited to integer values, the ratio of the LTPD to the AQL is also limited to certain values. For a fixed AQL, only certain LTPDs are possible. For example, when the AQL is 1.0%, an LTPD of 36.9% is possible as is 11.6%. However, intermediate values are not possible. Barely missing one of these ratios, results in the accept number jumping up to the next value, dramatically increasing the sample size. If you use the program SINGLE to select a sampling plan, you should compare the sampling plan you select with the closest match in Table 3.3. You may find that by making slight adjustments to the desired AQL or LTPD, you can dramatically reduce the sample size.

Table 3.3: Single Sampling Plans for Defectives by AQL and LTPD

AQL	Approximate Ratio of $LTPD/AQL$								
	45	11	6.5	5	4	3.2	2.8	2.3	2
10%	-	4/(1,2) AQL = 9.76 LTPD = 68.0 AOQL = 19.8	9/(2,3) AQL = 9.77 LTPD = 49.0 AOQL = 15.1	14/(3,4) AQL = 10.4 LTPD = 41.7 AOQL = 14.0	20/(4,5) AQL = 10.4 LTPD = 36.1 AOQL = 12.9	32/(6,7) AQL = 10.7 LTPD = 30.6 AOQL = 12.1	50/(8,9) AQL = 9.72 LTPD = 24.7 AOQL = 10.5	80/(12,13) AQL = 9.89 LTPD = 21.4 AOQL = 10.1	125/(18,19) AQL = 10.2 LTPD = 19.3 AOQL = 10.1
6.5%	-	6/(1,2) AQL = 6.28 LTPD = 51.0 AOQL = 13.4	13/(2,3) AQL = 6.61 LTPD = 36.0 AOQL = 10.5	20/(3,4) AQL = 7.13 LTPD = 30.4 AOQL = 9.75	32/(4,5) AQL = 6.36 LTPD = 23.4 AOQL = 8.00	50/(6,7) AQL = 6.76 LTPD = 20.1 AOQL = 7.70	80/(8,9) AQL = 6.00 LTPD = 15.7 AOQL = 6.49	125/(12,13) AQL = 6.26 LTPD = 13.9 AOQL = 6.42	200/(18,19) AQL = 6.31 LTPD = 12.2 AOQL = 6.25
4.0%	-	9/(1,2) AQL = 4.10 LTPD = 36.8 AOQL = 9.05	20/(2,3) AQL = 4.22 LTPD = 24.5 AOQL = 6.82	32/(3,4) AQL = 4.38 LTPD = 19.7 AOQL = 6.08	50/(4,5) AQL = 4.02 LTPD = 15.4 AOQL = 5.11	80/(6,7) AQL = 4.18 LTPD = 12.8 AOQL = 4.79	125/(8,9) AQL = 3.81 LTPD = 10.2 AOQL = 4.14	200/(12,13) AQL = 3.89 LTPD = 8.76 AOQL = 4.00	315/(18,19) AQL = 3.99 LTPD = 7.77 AOQL = 3.95
2.5%	2/(0,1) AQL = 2.53 LTPD = 68.4 AOQL = 14.8	13/(1,2) AQL = 2.81 LTPD = 26.8 AOQL = 6.32	32/(2,3) AQL = 2.60 LTPD = 15.8 AOQL = 4.27	50/(3,4) AQL = 2.78 LTPD = 12.9 AOQL = 3.89	80/(4,5) AQL = 2.49 LTPD = 9.74 AOQL = 3.19	125/(6,7) AQL = 2.66 LTPD = 8.27 AOQL = 3.06	200/(8,9) AQL = 2.37 LTPD = 6.42 AOQL = 2.58	315/(12,13) AQL = 2.46 LTPD = 5.59 AOQL = 2.53	500/(18,19) AQL = 2.50 LTPD = 4.92 AOQL = 2.48
1.5%	3/(0,1) AQL = 1.70 LTPD = 53.6 AOQL = 10.5	20/(1,2) AQL = 1.81 LTPD = 18.1 AOQL = 4.14	50/(2,3) AQL = 1.66 LTPD = 10.4 AOQL = 2.74	80/(3,4) AQL = 1.73 LTPD = 8.16 AOQL = 2.43	125/(4,5) AQL = 1.59 LTPD = 6.29 AOQL = 2.04	200/(6,7) AQL = 1.65 LTPD = 5.21 AOQL = 1.91	315/(8,9) AQL = 1.50 LTPD = 4.09 AOQL = 1.64	500/(12,13) AQL = 1.54 LTPD = 3.54 AOQL = 1.59	800/(18,19) AQL = 1.56 LTPD = 3.08 AOQL = 1.55
1.0%	5/(0,1) AQL = 1.02 LTPD = 36.9 AOQL = 6.70	32/(1,2) AQL = 1.12 LTPD = 11.6 AOQL = 2.60	80/(2,3) AQL = 1.03 LTPD = 6.52 AOQL = 1.71	125/(3,4) AQL = 1.10 LTPD = 5.27 AOQL = 1.55	200/(4,5) AQL = 0.990 LTPD = 3.96 AOQL = 1.27	315/(6,7) AQL = 1.05 LTPD = 3.32 AOQL = 1.21	500/(8,9) AQL = 0.942 LTPD = 2.59 AOQL = 1.03	800/(12,13) AQL = 0.964 LTPD = 2.21 AOQL = 0.995	1250/(18,19) AQL = 0.998 LTPD = 1.98 AOQL = 0.991
0.65%	8/(0,1) AQL = 0.639 LTPD = 25.0 AOQL = 4.33	50/(1,2) AQL = 0.715 LTPD = 7.56 AOQL = 1.67	125/(2,3) AQL = 0.657 LTPD = 4.20 AOQL = 1.10	200/(3,4) AQL = 0.686 LTPD = 3.31 AOQL = 0.971	315/(4,5) AQL = 0.627 LTPD = 2.52 AOQL = 0.808	500/(6,7) AQL = 0.659 LTPD = 2.10 AOQL = 0.763	800/(8,9) AQL = 0.588 LTPD = 1.62 AOQL = 0.715	1250/(12,13) AQL = 0.616 LTPD = 1.42 AOQL = 0.636	2000/(18,19) AQL = 0.623 LTPD = 1.24 AOQL = 0.619
0.4%	13/(0,1) AQL = 0.394 LTPD = 16.2 AOQL = 2.73	80/(1,2) AQL = 0.446 LTPD = 4.78 AOQL = 1.05	200/(2,3) AQL = 0.410 LTPD = 2.64 AOQL = 0.685	315/(3,4) AQL = 0.435 LTPD = 2.11 AOQL = 0.617	500/(4,5) AQL = 0.395 LTPD = 1.59 AOQL = 0.509	800/(6,7) AQL = 0.411 LTPD = 1.31 AOQL = 0.477	1250/(8,9) AQL = 0.376 LTPD = 1.04 AOQL = 0.412	2000/(12,13) AQL = 0.385 LTPD = 0.888 AOQL = 0.398	-
0.25%	20/(0,1) AQL = 0.256 LTPD = 10.9 AOQL = 1.79	125/(1,2) AQL = 0.285 LTPD = 3.08 AOQL = 0.670	315/(2,3) AQL = 0.260 LTPD = 1.68 AOQL = 0.435	500/(3,4) AQL = 0.274 LTPD = 1.33 AOQL = 0.388	800/(4,5) AQL = 0.247 LTPD = 0.997 AOQL = 0.318	1250/(6,7) AQL = 0.263 LTPD = 0.841 AOQL = 0.305	2000/(8,9) AQL = 0.235 LTPD = 0.649 AOQL = 0.257	-	-
0.15%	32/(0,1) AQL = 0.160 LTPD = 6.94 AOQL = 1.13	200/(1,2) AQL = 0.178 LTPD = 1.93 AOQL = 0.419	500/(2,3) AQL = 0.164 LTPD = 1.06 AOQL = 0.274	800/(3,4) AQL = 0.171 LTPD = 0.833 AOQL = 0.243	1250/(4,5) AQL = 0.158 LTPD = 0.638 AOQL = 0.204	2000/(6,7) AQL = 0.164 LTPD = 0.526 AOQL = 0.191	-	-	-

Table 3.3: (continued)

AQL	Approximate Ratio of $^{LTPD}/_{AQL}$								
	45	11	6.5	5	4	3.2	2.8	2.3	2
0.1%	50/(0,1) AQL = 0.103 LTPD = 4.50 AOQL = 0.728	315/(1,2) AQL = 0.113 LTPD = 1.23 AOQL = 0.266	800/(2,3) AQL = 0.102 LTPD = 0.664 AOQL = 0.171	1250/(3,4) AQL = 0.109 LTPD = 0.534 AOQL = 0.155	2000/(4,5) AQL = 0.0986 LTPD = 0.399 AOQL = 0.127	-	-	-	-
0.065%	80/(0,1) AQL = 0.0641 LTPD = 2.84 AOQL = 0.457	500/(1,2) AQL = 0.0711 LTPD = 0.776 AOQL = 0.168	1250/(2,3) AQL = 0.0655 LTPD = 0.425 AOQL = 0.110	2000/(3,4) AQL = 0.0683 LTPD = 0.334 AOQL = 0.0971	-	-	-	-	-
0.04%	125/(0,1) AQL = 0.0411 LTPD = 1.83 AOQL = 0.293	800/(1,2) AQL = 0.0444 LTPD = 0.485 AOQL = 0.105	2000/(2,3) AQL = 0.0408 LTPD = 0.266 AOQL = 0.0686	-	-	-	-	-	-
0.025%	200/(0,1) AQL = 0.0256 LTPD = 1.14 AOQL = 0.183	1250/(1,2) AQL = 0.0285 LTPD = 0.311 AOQL = 0.0672	-	-	-	-	-	-	-
0.015%	315/(0,1) AQL = 0.0103 LTPD = 0.728 AOQL = 0.117	2000/(1,2) AQL = 0.0178 LTPD = 0.194 AOQL = 0.0420	-	-	-	-	-	-	-
0.01%	500/(0,1) AQL = 0.0103 LTPD = 0.459 AOQL = 0.0735	-	-	-	-	-	-	-	-
0.0065%	800/(0,1) AQL = 0.00644 LTPD = 0.287 AOQL = 0.0460	-	-	-	-	-	-	-	-
0.004%	1250/(0,1) AQL = 0.00415 LTPD = 0.184 AOQL = 0.0294	-	-	-	-	-	-	-	-
0.0025%	2000/(0,1) AQL = 0.00253 LTPD = 0.115 AOQL = 0.0184	-	-	-	-	-	-	-	-

When defects are tallied, Table 3.4 should be used instead. To determine the single sampling plan with a specified AQL and LTPD, follow the procedure below:

(1) Calculate the ratio of $^{LTPD}/_{AQL}$.

(2) Find the largest ratio in Table 3.4 less than or equal to the calculated ratio. The corresponding accept number is given in the first column.

(3) Use the corresponding formulas to calculate the minimum and maximum sample sizes. Any sample size in this range may be used.

For example, assume one wants to select a single sampling plan with an AQL of 0.01 defects per unit and an LTPD of 0.1 defects per unit. The ratio is 10. The largest ratio in Table 3.4 less than or equal to 10 is 6.509. The corresponding accept number is 2. The minimum and maximum sample sizes are:

$$n_{min} = {}^{5.322}/_{LTPD} = 53.22 \qquad n_{max} = {}^{0.8177}/_{AQL} = 81.77$$

The smallest possible integer sample size is 54. Picking the smallest possible sample size causes the actual LTPD to closely match the specified LTPD. To match the AQL instead, choose n=81.

**Table 3.4: Constants for Selecting Single Sampling Plans
for Defects With a Specified AQL and LTPD**

Accept Number	Ratio of $^{LTPD}/_{AQL}$	Minimum Sample Size	Maximum Sample Size
0	44.89	$^{2.303}/_{LTPD}$	$^{0.05129}/_{AQL}$
1	10.94	$^{3.890}/_{LTPD}$	$^{0.3554}/_{AQL}$
2	6.509	$^{5.322}/_{LTPD}$	$^{0.8177}/_{AQL}$
3	4.890	$^{6.681}/_{LTPD}$	$^{1.366}/_{AQL}$
4	4.057	$^{7.993}/_{LTPD}$	$^{1.970}/_{AQL}$
6	3.206	$^{10.53}/_{LTPD}$	$^{3.285}/_{AQL}$
8	2.768	$^{12.99}/_{LTPD}$	$^{4.695}/_{AQL}$
12	2.314	$^{17.78}/_{LTPD}$	$^{7.690}/_{AQL}$
18	1.990	$^{24.76}/_{LTPD}$	$^{12.44}/_{AQL}$

Exercises

(3.1)   Assume defectives are tallied. Select a single sampling plan with an AQL of 0.4% and LTPD of 5% ensuring the specified protection for all lots sizes. Use both the table and program.

(3.2)   Assume defectives are tallied. Select a single sampling plan with an AQL of 0.4% and LTPD of 5% ensuring the specified protection for a lot of 500 units.

(3.3)   Assume defects are tallied and that the lot contains units of product. Select a single sampling plan with an AQL of 0.004 and LTPD of 0.05 ensuring the desired protection for all lot sizes. Use both the table and program.

(3.4)   Assume defects are tallied and that the lot contains units of product. Select a single sampling plan with an AQL of 0.004 and LTPD of 0.05 ensuring the desired protection for a lot of 500 units.

(3.5)   Assume defects are tallied, the lot is a continuous, and that the smallest sample size increment is 5. Select a single sampling plan with an AQL of 0.004 and LTPD of 0.05 ensuring the desired protection for all lot sizes. Use both the table and program.

3.2   Selecting Plans Based on AQL and AOQL

When customers receive a series of lots, they are probably more concerned with the overall level of defects than with the quality of individual lots. A second use of acceptance sampling is to assure the average outgoing quality is below some specified value. To provide this assurance, rejected lots must be 100% inspected.

When using acceptance sampling to guarantee the average outgoing quality, one needs to specify the desired value for the AOQL. The sampling plan will then ensure that the average outgoing quality is no greater than this AOQL. Once again, good lots should be routinely released and the sample size should be as small as possible. This can be accomplished by specifying an AQL above the historical process average and then selecting the sampling plan with the specified AQL and AOQL that minimizes the sample size (See Figure 3.3).

> (1)   Specify AOQL representing the maximum allowable level of defects or defectives for a long term average.
>
> (2)   Specify AQL above the historical process average.
>
> (3)   Select sampling plan minimizing the sample size where:
>
> $$actual\ AQL\ \geq\ specified\ AQL$$
>
> $$actual\ AOQL\ \leq\ specified\ AOQL$$

**Figure 3.3: Procedure for Selecting Sampling Plans
Guaranteeing the Average Outgoing Quality**

Following this procedure ensures:

**The quality of the product going to the
customer will average no worse than the AOQL.**

Once values for the AQL and AOQL have been selected, the program SINGLE can be used to select the single sampling plan satisfying the above conditions that minimizes the sample size. Begin by starting the program and selecting option 7 from the main menu. The program then requests the AQL and AOQL as well as additional information such as whether defects or defectives are tallied. The algorithm used for selecting single sampling plans for a specified AQL and AOQL is a slight modification of the algorithm used to select single sampling plans for a specified AQL and LTPD.

The AOQLs calculated by SINGLE as well as those given in the tables assume that the 100% inspection is 100% effective, that the lot size is large when compared to the sample size, and that defective units found in the sample or as part of the 100% inspection are replaced or reworked. Violation of the last two assumptions generally does not effect the results significantly. However, violation of the assumption that the 100% inspection is 100% effective can have a major effect. Section 2.5 showed the results of violating this assumption. As a rule of thumb, if the efficiency of the 100% inspection is at least 100-AOQL, then one need not be concerned.

As an example, suppose one wants to select a single sampling plan with an AQL of 1.0% and AOQL of 2.5% where the number of defectives is tallied. Screen 3.4 shows the required input. Pressing the enter key will cause the plan to be selected and displayed as in Screen 3.5.

Screen 3.4: Selection of Single Sampling Plan
Based on AQL and AOQL (SINGLE)

```
**************************************************************************
***                                                                  ***
***       SINGLE Option 7 - Select Sampling Plan Based on AQL and AOQL   ***
***                                                                  ***
**************************************************************************

    This option selects the single sampling plan minimizing the ASN curve
    from among those satisfying the following set of conditions:

        Actual AQL is greater than or equal to the specified AQL
        Actual AOQL is less than or equal to the specified AOQL

    The standard definition of AQL (95% chance of acceptance) is used unless
    indicated otherwise.  One must specify whether defects or defectives are
    tallied and, for defects, whether the lot is continuous or contains units.

    It is assumed that rejected lots are 100% inspected, the 100% inspection
    finds all defective units, defectives found by the inspection are repaired
    or replaced, and the lot size is large compared  to the sample size.

TALLY? (1=defectives,2=defects)              --> 1
ENTER AQL  AS PERCENT DEFECTIVE              --> 1
ENTER AOQL AS PERCENT DEFECTIVE             --> 2.5
USE .95 FOR AQL? ("y" or "n")               --> y
```

Screen 3.5: Result of Selecting Single Sampling Plan with
AQL=1%, AOQL=2.5%, Tally Defectives (SINGLE)

```
**************************************************************************
*                         *                                           *
*                         *       Copyright (C) 1992 Taylor Enterprises  *
*     Program SINGLE      *       P.O. Box 820, Lake Villa, IL 60046     *
*                         *              (708) 356-1074                  *
*                         *                                           *
**************************************************************************

            Current Sampling Plan:  n =          34,   a =      1
                                    tally defectives


                      -----  MENU OPTIONS  -----

            (1) Enter sampling plan
            (2) Summary information
            (3) OC curve
            (4) AOQ curve
            (5) ASN curve
            (6) Select sampling plan based on AQL and LTPD
            (7) Select sampling plan based on AQL and AOQL
            (8) Select sampling plan based on AQL and n


ENTER NUMBER OF OPTION OR "q" TO QUIT  -->
```

The selected single sampling plan is n=34 and a=1. Using option 2, this plan has an actual AQL of 1.06% and an actual AOQL of 2.45%. This satisfies the required conditions that:

Actual AQL (1.06%) $\geq$ Specified AQL (1%)

Actual AOQL (2.45%) $\leq$ Specified AOQL (2.5%)

Of these plans, the selected plan minimizes the sample size.

Table 3.3 (page 47) can also be used to look up sampling plans with a specified AQL and AOQL when defectives are tallied. First find the row corresponding to an AQL of 1%. Go across this row until one finds the desired AOQL. In this case, the single sampling plan n=32 and a=1 is selected.

The program SINGLE can also be used when defects are tallied. In this case, one must specify whether the lot contains units or is continuous. In the latter case, the minimum sample size increment must also be entered. An alternative is to use Table 3.5 as follows:

(1) Calculate the ratio of $^{AOQL}/_{AQL}$.

(2) Find the largest ratio in Table 3.5 less than or equal to the calculated ratio. The corresponding accept number is given in the first column.

(3) Use the corresponding formulas to calculate the minimum and maximum sample sizes. Any sample size in this range may be used.

For example, assume one wants to select a single sampling plan with an AQL of 0.01 and AOQL of 0.025 defects per unit. The ratio is 2.5. The largest ratio in Table 3.5 less than or equal to the calculated ratio is 2.364. The corresponding accept number is 1. The minimum and maximum sample sizes are:

$$n_{min} = {}^{0.8400}/_{AOQL} = 33.6 \qquad n_{max} = {}^{0.3554}/_{AQL} = 35.54$$

The smallest integer sample size is 34.

Selecting sampling plans based on AOQLs can result in unnecessarily tight sampling plans. This is because the AOQL represents the worse possible performance of the sampling plan. The conditions resulting in this worse case performance may be unlikely to occur. Under typical conditions, the average outgoing quality is likely to be considerable better than the AOQL. The AOQL should be used only if nothing is known about the behavior of the incoming quality. If the behavior of the incoming quality is known, a better approach is to maximize the customer value. This is the topic of the next chapter.

Table 3.5: Constants for Selecting Single Sampling Plans
for Defects With a Specified AQL and AOQL

Accept Number	Ratio of AOQL/AQL	Minimum Sample Size	Maximum Sample Size
0	7.172	$0.3679 / AOQL$	$0.05129 / AQL$
1	2.364	$0.8400 / AOQL$	$0.3554 / AQL$
2	1.677	$1.371 / AOQL$	$0.8177 / AQL$
3	1.422	$1.942 / AOQL$	$1.366 / AQL$
4	1.291	$2.543 / AOQL$	$1.970 / AQL$
6	1.160	$3.812 / AOQL$	$3.285 / AQL$
8	1.096	$5.146 / AOQL$	$4.695 / AQL$
12	1.034	$7.948 / AOQL$	$7.690 / AQL$
18	0.9945	$12.37 / AOQL$	$12.44 / AQL$

Exercises

(3.6)   Assume that defectives are tallied. Select a single sampling plan with an AQL=0.4% and AOQL=1%. Use both the program and table.

(3.7)   Assume that defects are tallied and that the lot contains units. Select a single sampling plan with an AQL=0.004 and AOQL=0.01. Use both the program and table.

(3.8)   Assume defects are tallied and that the lot is continuous with a smallest sample size increment of 10. Select a single sampling plan with an AQL=0.004 and AOQL=0.01. Use both the program and table.

3.3   Selecting Plans to Reject the Obvious

Once process quality improves to the point that the process consistently produces good lots, sampling inspection can no longer be economically justified. All lots should be released. However, even a process that consistently produces good lots runs the risk of developing a problem sometime in the future. One cannot simply walk away from such a process. Some sort of control or monitoring procedure should be used. These both require the routine inspection of samples from the process. Whenever samples are inspected, regardless of the reason, it is possible for one of these samples to provide clear evidence that a lot is of poor quality. Provisions should be made for the rejection of such a lot. It would be hard to defend the release of a lot where there is clear evidence to the contrary.

To ensure obviously bad lots are rejected, accept numbers should be determined for all samples selected, regardless of their purpose. The accept number should be selected to ensure that normal production is routinely released. This can be accomplished by specifying an AQL above the historical process average. The procedure for selecting single sampling plans to reject obviously bad lots is shown in Figure 3.4. Rejection by this sampling plan provides 95% confidence that the lot's quality exceeds the sampling plan's AQL, i.e., that the lot is not representative of normal production.

---

(1)   Determine sample size of control or monitoring procedure.

(2)   Specify AQL above the historical process average.

(3)   Select sampling plan minimizing the LTPD where:

actual AQL $\geq$ specified AQL

n = sample size of control or monitoring procedure

**Figure 3.4: Procedure for Selecting Sampling Plans
For Rejecting Obviously Bad Lots**

Option 8 of program SINGLE can be used to select the sampling plan with a specified sample size and AQL which minimizes the LTPD. For example, suppose 15 samples are routinely selected and that an AQL of 1.0% is desired. Then Screen 3.6 shows the required input assuming defectives are tallied. Pressing the enter key results in the sampling plan being selected and displayed as in Screen 3.7. The selected sampling plan has n=15 and a=1.

Screen 3.6: Screen for Selecting Single Sampling Plans Based on Sample Size and AQL (SINGLE)

```
********************************************************************************
***                                                                        ***
***         SINGLE Option 8 - Select Sampling Plan Based on AQL and n       ***
***                                                                        ***
********************************************************************************

   This option selects the single sampling plan with the specified sample
   size and the smallest possible accept number (LTPD,AOQL) satisfying:

      Actual AQL is greater than or equal to the specified AQL

   The standard definition of AQL (95% chance of acceptance) is used
   unless indicated otherwise.  One must specify whether defects or
   defectives are tallied.  Type-B OC curves are assumed.

TALLY? (1=defectives,2=defects)            --> 1
ENTER SAMPLE SIZE    (n)                    --> 15
ENTER AQL  AS PERCENT DEFECTIVE            --> 1
USE .95 FOR AQL? ("y" or "n")             --> y
```

Screen 3.7: Result of Selecting Single Sampling Plan with AQL=1%, n=15, Defectives Tallied (SINGLE)

```
********************************************************************************
*                              *                                              *
*                              *      Copyright (C) 1992 Taylor Enterprises   *
*     Program SINGLE           *         P.O. Box 820, Lake Villa, IL 60046   *
*                              *                 (708) 356-1074               *
*                              *                                              *
********************************************************************************

          Current Sampling Plan:   n =        15,    a =     1
                                   tally defectives


                    ----- MENU OPTIONS  -----

          (1) Enter sampling plan
          (2) Summary information
          (3) OC curve
          (4) AOQ curve
          (5) ASN curve
          (6) Select sampling plan based on AQL and LTPD
          (7) Select sampling plan based on AQL and AOQL
          (8) Select sampling plan based on AQL and n


ENTER NUMBER OF OPTION OR "q" TO QUIT  -->
```

Similarly, plans can be selected when defects are tallied. In this case it must be specified whether the lot contains units or is continuous. The sampling plans selected are always based on the Type-B OC curve.

The protection afforded by a sampling plan selected to reject the obvious may not be much. However, in the event of a broken tool, clogged filter, or other break down of the process, 50% to 100% of the product may suddenly be defective. It is important that at least some samples be selected to detect such an event. It is also important that in the event the monitoring of the process detects an obviously bad lot, procedures be in place to reject it.

Exercises

(3.9)   Assume that a sample of 5 units is routinely inspected to control the process. If defectives are tallied and an AQL of 1% is desired, what sampling plan should be used to reject obviously bad lots.

(3.10) Assume that a sample of 5 units is routinely inspected to monitor process performance. If defects are tallied, the lot contains units of product, and an AQL of 0.01 defects per unit is desired, what sampling plan should be used to reject obviously bad lots.

3.4   Summary

There are four basic strategies for selecting sampling plans: as an insurance policy against highly defective lots, to guarantee the average outgoing quality, to reject obviously bad lots, and to maximize customer value. The first three strategies are covered in this chapter. The fourth strategy is the topic of chapter 4.

Selecting sampling plans to protect against highly defective lots involves specifying the AQL and LTPD and minimizing the sample size. One can state that with 90% confidence accepted lots are better than the LTPD. Since this approach specifies the protection on a lot by lot basis, a Type-A OC curve should be used. Specifying a Type-B OC curve instead, results in a sampling plan providing the desired protection regardless of the lot size. However, as a result, the sample size may be larger than necessary. Single sampling plans can be selected to protect against highly defective lots using the program SINGLE or the tables provided.

When rejected lots are 100% inspected, sampling plans can also be selected to guarantee the AOQL is below a specified value. Selecting sampling plans to ensure the average outgoing quality involves specifying the AQL and AOQL and minimizing the sample size. Such sampling plans guarantee the average outgoing quality is no worse than the specified AOQL. This approach assumes that the 100% inspection is 100% effective, defectives found as part of the inspection are replaced or reworked, and that the lot size is large compared to the resulting sample size. Of these assumptions, the assumption of 100% effective is most problematic. Single sampling plans can be selected to protect against highly defective lots using the program SINGLE or the tables provided.

A third use is to reject obviously bad lots. Anytime units are inspected, regardless of the reason, a procedure should be in place to reject obviously bad lots. This approach requires selecting the single sampling plan with matching sample size and specified AQL that minimizes the LTPD. With 95% confidence, lots rejected by this sampling plan are worse than the AQL, i.e., are different from normal production. Single sampling plans can be selected to reject obviously bad lots using the program SINGLE.

References

Guenther, William C.. (1977). *Sampling Inspection in Statistical Quality Control.* MacMillan, New York.

Taylor, Wayne A. (1983). *Selecting Efficient Single Stage and Double Stage Attribute Sampling Plans of a Given Power.* Ph.D. Thesis, Purdue University.

Economical
Single Sampling Plans

Single sampling plans can also be selected based on economics. This approach balances the cost of inspection with the cost of production and the cost of releasing defective units. The objective is to select the sampling plan or other action that maximizes customer value. One has a choice of five different actions:

- Release all lots

- Discard all lots

- 100% inspect all lots

- Acceptance sample, discarding rejected lots

- Acceptance sample, 100% inspecting rejected lots

Further, defectives found as part of a 100% inspection can either be discarded or reworked.

This chapter shows how to identify the action maximizing customer value. This chapter's objectives are:

- Learn to select the single sampling plan maximizing customer value when rejected lots are discarded.

- Learn to select the single sampling plan maximizing customer value when rejected lots are 100% inspected.

- Learn to compare the customer value of these sampling plans with the customer value for the other actions in order to select the best action.

4.1   An Example

Acceptance sampling is used to improve the quality of the product going to the customer. Ideally, all units sent to the customer should be defect free. However, many, if not all processes, produce some defects. Decisions must be made as to what to do with these lots. The action most consistent with the goal of defect-free product is to 100% inspect all lots. However, when testing is destructive, 100% inspection is not possible. Further, even when 100% inspection is possible, it can be expensive. Frequently, the cost of performing a 100% inspection exceeds its benefits.

Suppose it costs $500 to 100% inspect a lot and that the combined losses incurred by the customer and the company each time the customer receives a defective unit is $100. If only one defective unit is found as a result of the 100% inspection of a lot, the $500 spent to inspect the lot results in only a $100 return. The net loss of $400 must be passed on to the customer in the form of higher prices. While the 100% inspection resulted in better quality, it also resulted in lower customer value. One is not doing the customer a favor by spending $500 dollars to obtain $100 in return. The best action depends on the process quality and the costs involved. This chapter shows how to determine the best action. An example is given first.

A switch used in television sets is currently 100% inspected. Quality improvement efforts have recently reduced the incidence of defective switches. As a result, the elimination of the 100% inspection is being considered. To address this issue, one needs to know the lot size, the quality of the lots, and the costs involved. These are as follows:

U_p =	Average lot percent defective	= 1%
S_p =	Standard deviation of lot percent defective	= 2%
N =	Lot size	= 1000
P =	Price unit sold for	= $5
D =	Discard value of unit	= $0
L =	Loss incurred per defective unit sold	= $50
C =	Per unit cost of manufacturing	= $2
U_h =	Per unit cost of the 100% inspection	= $0.25
S_h =	Per lot setup cost of the 100% inspection	= $100
E =	Efficiency of the 100% inspection	= 90%
R =	Per unit cost of rework	not possible
U_s =	Per unit cost of acceptance sampling	= $0.50
S_s =	Per lot setup cost of acceptance sampling	= $0

These parameters are covered in detail in Sections 4.3 and 4.4.

The action maximizing customer value is desired. Customer value is:

$$
\begin{matrix}
\text{Customer} \\
\text{Value}
\end{matrix}
=
\begin{matrix}
\text{Income From} \\
\text{Sale of Units} \\
\text{and Discards}
\end{matrix}
-
\begin{matrix}
\text{Production} \\
\text{Costs}
\end{matrix}
-
\begin{matrix}
\text{Inspection} \\
\text{Costs}
\end{matrix}
-
\begin{matrix}
\text{Loss Due to} \\
\text{Sale of} \\
\text{Defectives}
\end{matrix}
$$

Throughout this book, whenever customer value is reported, it is reported on a per lot basis. To obtain customer value on a per unit basis, one should divide by the lot size. Customer value depends on the action taken, the costs involved, and the process quality. For the television switch, Figure 4.1 shows plots of customer value versus process quality for the actions release, 100% inspect, and discard.

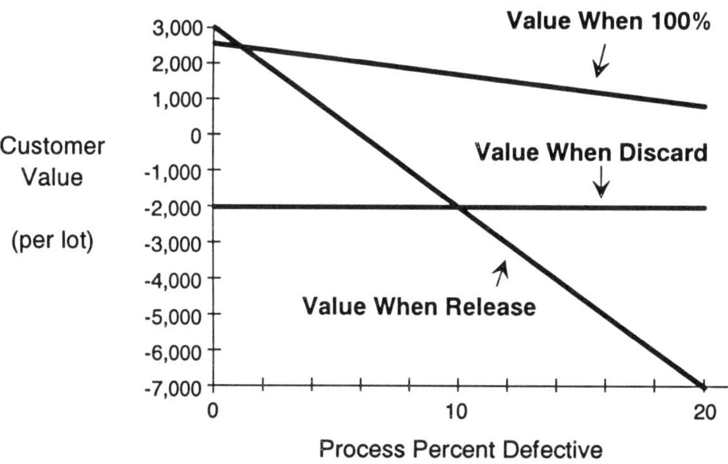

**Figure 4.1: Customer Value Versus Process Quality
For the Television Switch**

Of particular interest is the point where the lines "Value When Release" and "Value When 100%" cross. They cross at 0.864% defective. This point is the break even quality between releasing lots and 100% inspection. Lots below 0.864% defective should be released. Lots above 0.864% defective should be 100% inspected. The lines "Value When Release" and "Value When Discard" cross at 10% defective. This is the break even quality betwen releasing lots and discarding lots. The lines "Value When 100%" and "Value When Discard" also cross off the right of the plot. They cross at 48.9% defective. Based on Figure 4.1, lots below 0.864% defective should be released, lots between 0.864% and 48.9% defective should be 100% inspected and lots above 48.9% defective should be discarded.

The action maximizing customer value depends on the process quality. Selecting the best action requires that information on the process quality be obtained. The two parameters U_p and S_p are used to describe process quality. U_p is the average percent defective. S_p describes how much individual lots vary around U_p. Figure 4.2 shows the distribution of lot quality when U_p = 1% and S_p = 2%. Figure 4.2 can be thought of as a smoothed out histogram. The height of the curve above a value represents the frequency of that value. A procedure for estimating U_p and S_p is given in Section 4.4.

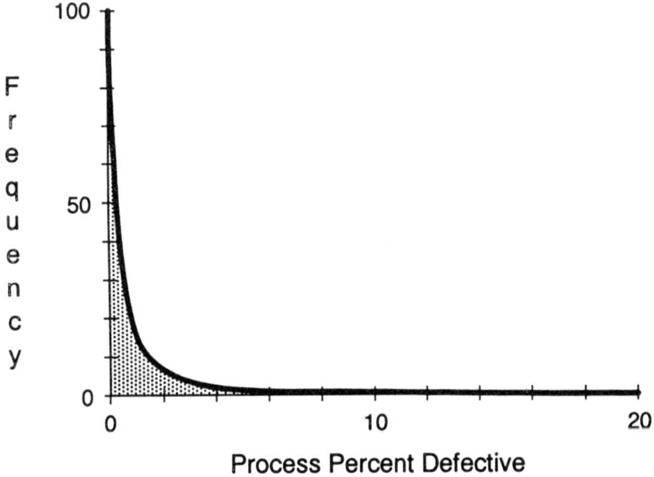

Figure 4.2: Distribution of Process Quality for U_p = 1% and S_p = 2%

The break even quality between releasing lots and 100% inspection is 0.864% defective. From Figure 4.2, approximately 72% of the lots fall below 0.864% defective and are best released. The other 28% of the lots fall above 0.864% and are best 100% inspected. A sampling plan that accepts most lots better than 0.864%, and rejects most lots above 0.864% could out perform either of these two actions, so long as the cost of inspection is not too high. When rejected lots are 100% inspected, the single sampling plan maximizing customer value for the distribution of quality in Figure 4.2 is n=80 and a=0. When rejected lots are discarded, the single sampling plan maximizing customer value for the distribution of quality in Figure 4.2 is n=57 and a=7. Table 4.1 shows the resulting customer values for the five possible actions. The best choice is to use the sampling plan n=80 and a=0 and to 100% inspect rejected lots. Use of this sampling plan saves close to $200 per lot compared to the next best choice.

Table 4.1: Customer Value for 5 Possible Actions

Action	Customer Value
Release All Lots	$2500.00
Discard All Lots	$-2000.00
100% Inspect All Lots	$2555.00
Use n=52 and a=7. Discard Rejected Lots.	$2502.95
Use n=80 and a=0. 100% Inspect Rejected Lots.	$2738.28

Figure 4.3 shows the OC curve for the single sampling plan n=80 and a=0. It has an AQL of 0.164%, IQ of 0.863%, and LTPD of 2.84%. The indifference quality is close to the break even quality of 0.864%. Further, the OC curve is steep enough to reject a sizable percentage of the lots that should be 100% inspected, yet release most of the lots that should be released.

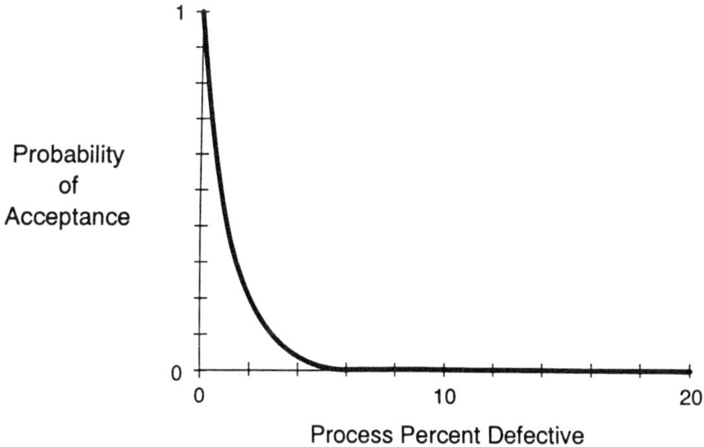

Figure 4.3: Type-B OC Curve of Sampling Plan n=80 and a=0

Figure 4.4 shows the relationship between customer value and process quality for this sampling plan. This sampling plan is almost as good as releasing all lots when releasing all lots is the best decision. It is almost as good as 100% inspecting all lots when 100% inspecting all lots is the best action. It provides the best of both these actions. The difference between the customer value of this sampling plan and that of the best action is the cost of inspection and the costs associated with misclassifying lots.

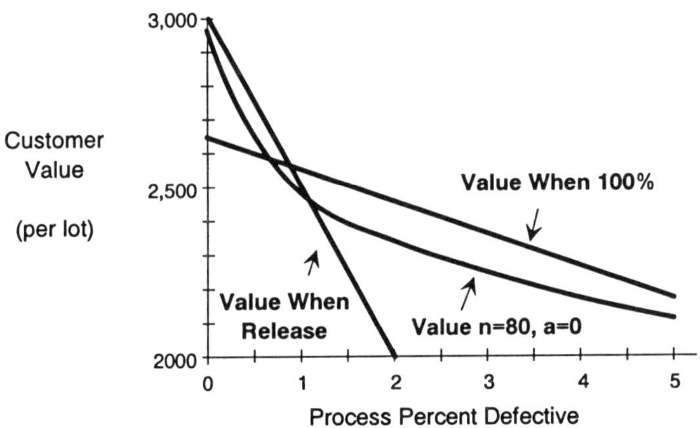

Figure 4.4: Customer Value of Single Sampling Plan n=80 and a=0 When Rejected Lots are 100% Inspected

4.2   Selecting the Action Maximizing Customer Value

This section shows how to select the sampling plan or other action maximizing customer value. In this section and Sections 4.3 to 4.5, it is assumed that defectives are tallied. Section 4.7 covers the tallying of defects.

Program VALUE on the Sampling Programs diskette can be used to select sampling plans maximizing customer value. It can also be used to compare the customer value of these sampling plans with those of the other actions to determine the best action. When the program VALUE is first started, the main menu shown in Screen 4.1 is displayed.

The first four options are used to enter the various parameters. Option 5 allows a sampling plan to be specified. Asterisks are displayed in front of each option for which information has been entered. The last three options are used to select the sampling plan maximizing customer value and to compare it with the other actions. An "x" is displayed in front of those analysis options that are currently disabled. They are initially disabled and remain disabled until all the information required to complete the analysis has been entered. The required data entry options are listed in parenthesis. Typically one enters the parameters using options 1 through 4, selects the best sampling plan using option 8, and then compares the sampling plan with the other actions using option 7.

Screen 4.1: Main Menu (VALUE)

```
***********************************************************************
*                         *                                          *
*                         *    Copyright (C) 1992 Taylor Enterprises *
*    Program VALUE        *       P.O. Box 820, Lake Villa, IL 60046 *
*                         *               (708) 356-1074             *
*                         *                                          *
***********************************************************************

                      -----  MENU OPTIONS  -----

          DATA ENTRY OPTIONS (* means completed):

                  (1) release-discard cost parameters
                  (2) 100% inspection cost parameters
                  (3) process quality parameters
                  (4) sampling inspection cost parameters
                  (5) single sampling plan

          ANALYSIS OPTIONS (x means further data required):

              x (6) plot or table of customer value (1)
              x (7) table of average values for actions (1,3)
              x (8) select sampling plan maximizing value (1,3,4)

ENTER NUMBER OF OPTION OR "q" TO QUIT  -->
```

Option 1 is used to view and input the release-discard cost
parameters: N, P, D, L and C. Screen 4.2 shows the input of these
parameters. Option 2 is used to view and input the 100% inspection
cost parameters: U_h, S_h, E and R. This option is only necessary if
testing is nondestructive. Screen 4.3 shows the input of these
parameters.

Screen 4.2: Entry of Release-Discard Cost Parameters (VALUE)

```
***********************************************************************
***                                                               ***
***         VALUE Option 1 - Enter Release-Discard Cost Parameters ***
***                                                               ***
***********************************************************************

   The release-discard cost parameters are:

      N = average lot size
      P = price units sold for (per unit)
      D = discard or salvage value of units (per unit)
      L = loss incurred per defective unit sold
      C = per unit cost of manufacturing (fully loaded)

ENTER NEW VALUES? ("y" or "n")        --> y
ENTER AVERAGE LOT SIZE (N)            --> 1000
ENTER PER UNIT SALES PRICE (P)        --> 5
ENTER PER UNIT DISCARD VALUE (D)      --> 0
ENTER LOSS PER DEFECTIVE SOLD (L)     --> 50
ENTER COST OF MANUFACTURING UNIT (C) --> 2
```

Screen 4.3: Entry of 100% Inspection Cost Parameters (VALUE)

```
**************************************************************************
***                                                                   ***
***          VALUE Option 2 - Enter 100% Inspection Cost Parameters    ***
***                                                                   ***
**************************************************************************

   The 100% inspection cost parameters are:

      UH = per unit cost of 100% inspection
      SH = per lot setup cost of 100% inspection
      E  = efficiency of 100% inspection (percent)
      R  = rework cost per unit

   If no values are entered, it is assumed 100% inspection is not possible.

ENTER NEW VALUES? ("y" or "n")              --> y
CAN UNITS BE 100% INSPECTED? ("y" or "n") --> y
ENTER PER UNIT COST (UH)                    --> .25
ENTER PER LOT SETUP COST (SH)               --> 100
ENTER EFFICIENCY (E)                        --> 90
CAN UNITS BE REWORKED? ("y" or "n")         --> n
```

At this point, option 6 can be used to obtain a plot or table of
customer value versus process quality for the actions release, discard,
and 100% inspect. Break even qualities are also given. When option
6 is selected, Screen 4.4 appears asking whether a plot or table is
desired and how it should be scaled. In Screen 4.4, a plot of all three
actions is requested ranging from 0% to 20% percent defective. The
default scale for customer value is used.   Screen 4.5 shows the
resulting plot. This plot is the same as the one in Figure 4.1.

Screen 4.4: Input for Plot of Customer Value (VALUE)

```
**************************************************************************
***                                                                   ***
***          VALUE Option 6 - Plot or Table of Customer Value          ***
***                                                                   ***
**************************************************************************

   This option gives a plot or table of the customer value versus
   process quality.  In either case one must enter the minimum and maximum
   percent defective.  For a table, one must also enter the increment.
   For a plot one can optionally specify a range for customer value.

ENTER "p" FOR PLOT "t" FOR TABLE.            --> p
ENTER MINIMUM PERCENT DEFECTIVE              --> 0
ENTER MAXIMUM PERCENT DEFECTIVE              --> 20
USE DEFAULT SCALE FOR VALUE? ("y" or "n")    --> y
PLOT ALL ACTIONS? ("y" or "n")               --> y
```

Screen 4.5: Plots of Customer Value (VALUE)

```
            PLOT OF CUSTOMER VALUE VERSUS PROCESS QUALITY
        $        3000.00 +rr
                         |hhhhhhhhhhhhh
     Customer Value      |     rrr        hhhhhhhhhhhhh
        (per lot)        |       rr                    hhhhhhhhhhhhh
                         |        rrr                               hh
  r = Release All Lots   |          rr
  d = Discard All Lots   |           rrr
  h = 100% Inspect (discard)  |          rr
                         +dddddddddddddddddddddddddddddddddddddddddddddd
                         |              rr
                         |               rrr
  Break Even Qualities:  |                 rr
    r/d =  10.000000%    |                  rrr
    r/h =    .864198%    |                    rr
    h/d =  48.947370%    |                     rrr
                         |                       rr
        $       -7000.00 +                        rr
                         +------------------+------------------+
                  .000000%                            20.000000%

                     Process Quality (percent defective)

PRESS ENTER KEY TO CONTINUE
```

Following the plot, additional screens appear displaying all the cost
and process parameters so that they can be printed out along with
the plot. If a sampling plan and its associated costs had also been
entered, its customer value would also have been plotted, both when
rejected lots are discarded and when rejected lots are 100% inspected.
When all five actions are plotted, the plot can become crowded. The
last question on the input screen allows one to select which actions to
plot.

Screen 4.6: Input for Table of Customer Value (VALUE)

```
**********************************************************************************
***                                                                          ***
***            VALUE Option 6 - Plot or Table of Customer Value              ***
***                                                                          ***
**********************************************************************************

   This option gives a plot or table of the customer value versus
   process quality.  In either case one must enter the minimum and maximum
   percent defective.  For a table, one must also enter the increment.
   For a plot one can optionally specify a range for customer value.

ENTER "p" FOR PLOT "t" FOR TABLE.                --> t
ENTER MINIMUM PERCENT DEFECTIVE                  --> 0
ENTER MAXIMUM PERCENT DEFECTIVE                  --> 20
ENTER INCREMENT PERCENT DEFECTIVE FOR TABLE --> 2
```

Screen 4.6 shows the use of option 6 to obtain a table. A range of 0% to 20% defective is specified with an increment of 2%. As a result, the table includes process percent defectives of 0%, 2%, 4%, ⋯, 20%. The resulting table is shown in Screen 4.7. It is followed by Screen 4.8 displaying the break even quantities. Following these two screens, the cost and process parameters are also displayed.

Screen 4.7: Table of Customer Value (VALUE)

```
                      CUSTOMER VALUE (per lot)

  PROCESS      RELEASE      DISCARD        100%        SAMPLE       SAMPLE
  PERCENT       ALL           ALL        INSPECT      DISCARD        100%
  DEFECTIVE    LOTS          LOTS       ALL LOTS     REJECTIONS    REJECTIONS
  ----------   ----------   ----------   ----------   ----------   ----------

    .000000     3000.00     -2000.00      2650.00
   2.000000     2000.00     -2000.00      2460.00
   4.000000     1000.00     -2000.00      2270.00
   6.000000       .00       -2000.00      2080.00
   8.000000    -1000.00     -2000.00      1890.00
   9.999999    -2000.00     -2000.00      1700.00
  12.000000    -3000.00     -2000.00      1510.00
  14.000000    -4000.00     -2000.00      1320.00
  16.000000    -5000.00     -2000.00      1130.00
  18.000000    -6000.00     -2000.00       940.00
  20.000000    -7000.00     -2000.00       750.00

PRESS ENTER KEY TO CONTINUE
```

Screen 4.8: Break Even Qualities and Notes (VALUE)

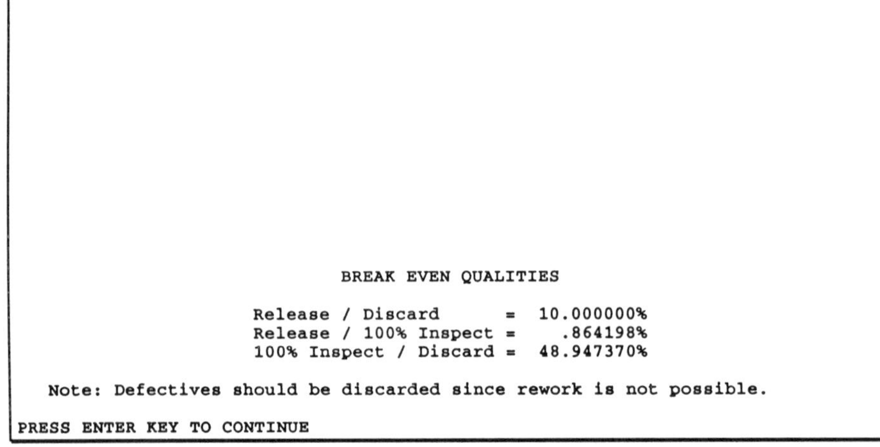

```
                      BREAK EVEN QUALITIES

           Release / Discard       =   10.000000%
           Release / 100% Inspect  =     .864198%
           100% Inspect / Discard  =   48.947370%

  Note: Defectives should be discarded since rework is not possible.

PRESS ENTER KEY TO CONTINUE
```

Rework is not possible for the television switch. As a result, Screens 4.5 and 4.8 indicate defective units found as part of the 100% inspection should be discarded. If rework were possible, these messages would change. Rework should only be performed if the cost of rework, R, is less than the resulting benefits. The benefit of rework is to increase a unit's value from its discard value D to its selling price P. Rework should only be performed if R < (P-D). In this case, these messages will indicate that defective units are to be reworked and all customer values and break even qualities displayed will make this assumption. If $R \geq$ (P-D), these messages will indicate that defectives should be discarded and this assumption will be made when displaying customer values and break even qualities.

Before any additional analysis can be performed, the two parameters describing process quality must be entered. For the television switch, historical data was used to estimate these parameters. The resulting estimates are U_p = 1% and S_p = 2%. Section 4.4 (page 76) gives the procedure used. Option 3 is used to view and enter this information.

Two different distributions for process quality can be entered. The first distribution describes normal operating conditions. The second distribution describes the distribution of quality under problem conditions. One must also enter the percentage of time that normal operating conditions apply. For the TV switch, it was assumed that normal operating conditions were in effect 100% of the time. As a result, no second distribution was entered. The entry of the process quality parameters is shown in Screen 4.9.

Screen 4.9: Entry of Process Quality Parameters (VALUE)

```
************************************************************************
***                                                                ***
***            VALUE Option 3 - Enter Process Quality Parameters    ***
***                                                                ***
************************************************************************

   The process quality parameters are:

      Normal Conditions:
         Up1 = average percent defective
         Sp1 = standard deviation of percent defective
         P1  = percentage time under normal conditions

      Problem Conditions:
         Up2 = average percent defective
         Sp2 = standard deviation of percent defective

ENTER NEW VALUES? ("y" or "n") --> y
ENTER Up1                      --> 1
ENTER Sp1                      --> 2
ENTER P1                       --> 100
```

The final set of parameters are the sampling inspection costs: U_s and S_s. These are viewed and entered using option 4. Screen 4.10 shows the input of these values.

Screen 4.10: Entry of Sampling Cost Parameters (VALUE)

```
****************************************************************************
***                                                                    ***
***        VALUE Option 4 - Enter Sampling Inspection Cost Parameters   ***
***                                                                    ***
****************************************************************************

   The sampling inspection cost parameters are:

      US = per unit cost of sampling inspection
      SS = per lot setup cost of sampling inspection

ENTER NEW VALUES? ("y" or "n")        --> y
ENTER PER UNIT COST (US)              --> .5
ENTER PER LOT SETUP COST (SS)         --> 0
```

The sampling plan maximizing customer value can now be selected using option 8. Screen 4.11 shows the results. Upon selecting option 8, one is asked whether to restrict the LTPD. A reply of no was given. Section 4.6 discusses restricted economic sampling plans.

Screen 4.11: Selecting Best Sampling Plan (VALUE)

```
****************************************************************************
***                                                                    ***
***        VALUE Option 8 - Select Most Economical Single Sampling Plan ***
***                                                                    ***
****************************************************************************

   When rejected lots are discarded, the most economical single
   sampling plan is:

      n    =        52
      a    =         7
      Value = $    2502.95      (per lot)

   When rejected lots are 100% inspected, the most economical single
   sampling plan is:

      n    =        80
      a    =         0
      Value = $    2738.28      (per lot)

   Rework is not possible so defective units should be discarded.

   The better plan is the one where rejected lots are 100% inspected.

PRESS ENTER KEY TO CONTINUE
```

Actually, two sampling plans are selected. The first plan, n=52 and a=7, maximizes customer value when rejected lots are discarded. The resulting value is $2502.95. The second sampling plan, n=80 and a=0, maximizes the value when rejected lots are 100% inspected. The resulting value, $2738.28, is greater than that of the first plan. Therefore, the second sampling plan is the better of the two.

All that remains is to compare the customer value of the selected sampling plan with those for the other actions. Option 7 is used for this purpose. The results are shown in Screens 4.12 and 4.13. The action maximizing customer value is the single sampling plan n=80 and a=0 where rejected lots are 100% inspected. This saves almost $200 per lot over the second choice of 100% inspecting all lots. The first table given by option 7 also shows the effect of each action on the outgoing quality. The incoming quality is 1% defective. Use of this sampling plan reduces the outgoing quality to 0.23% while only 100% inspecting 30% of the lots. This sampling plan offers 85% of the benefits of 100% inspection at only 30% of the effort. The second table given by option 7 breaks down customer value into its components.

While there is just one sampling plan maximizing customer value, there are many others nearly as good. Option 5 on the main menu allows one to explore the effect of using other single sampling plans.

Screen 4.12: Compare Actions - First Table (VALUE)

```
****************************************************************************
***                                                                    ***
***          VALUE Option 7 - Table of Customer Values (per lot)        ***
***                                                                    ***
****************************************************************************


                  AVERAGE        PERCENT        PERCENT
                  OUTGOING       OF LOTS        OF LOTS        CUSTOMER
      ACTION      QUALITY        DISCARDED      100% INSPEC    VALUE
   -----------    ----------     -----------    -----------    ----------

   Release All    1.0000000        .0000000        .0000000      2500.00
   Discard All                  100.0000000        .0000000     -2000.00
   100% All        .1000000        .0000000     100.0000000      2555.00
   Samp + Disc     .2106024      29.8855700        .0000000      1391.89
   Samp + 100%     .2253762        .0000000      29.8855700      2738.28


           The average incoming quality is =    1.0000000000%.
           Current Sampling Plan is n =             80 and a =       0.
           Defectives should be discarded since rework is not possible.


PRESS ENTER KEY TO CONTINUE
```

Screen 4.13: Compare Actions - Second Table (VALUE)

```
*******************************************************************************
***                                                                       ***
***             VALUE Option 7 - Table of Customer Values (per lot)        ***
***                                                                       ***
*******************************************************************************


               INCOME FROM
               SALES OF                LOSS DUE TO
               UNITS AND     COST OF     SALE OF      COST OF      CUSTOMER
    ACTION     DISCARDS    MANUFACTURE  DEFECTIVES   INSPECTION     VALUE
  ----------- -----------  -----------  -----------  -----------  -----------

Release All     5000.00 -    2000.00 -     500.00 -        .00 =    2500.00
Discard All         .00 -    2000.00 -        .00 -        .00 =   -2000.00
100% All        4955.00 -    2000.00 -      50.00 -     350.00 =    2555.00
Samp + Disc     3505.72 -    2000.00 -      73.83 -      40.00 =    1391.89
Samp + 100%     4989.35 -    2000.00 -     112.45 -     138.62 =    2738.28


               Current Sampling Plan is n =        80 and a =      0.
               Defectives should be discarded since rework is not possible.


PRESS ENTER KEY TO CONTINUE
```

Exercises

(4.1)   For the television switch, suppose that 100% inspection is not possible. Which action now maximizes customer value?

(4.2)   For the television switch, suppose defective units can be reworked at a cost of $2 each. Should defectives be reworked? How about if the cost of rework is $6 per unit?

(4.3)   One suggestion for reducing inspection costs is to increase the lot size. What action maximizes customer value if the lot size changes to 10,000? What is the per unit value of this action? Compare this to the best action when N=1000.   What assumption is made in estimating this savings?

(4.4)   An engineer proposes adding an automatic 100% inspection device. The error rate of the device is 20% (E=80%). The cost of implementation is $20,000.   Spreading this cost over 9 months, results in a per unit inspection cost U_h of $0.003. The per lot setup cost S_h is zero since the inspection is performed on-line. Should the proposal be implemented?

(4.5)   The sampling plan maximizing customer value is n=80 and a=0. If the process quality changes to $U_p = 0.5\%$ and $S_p = 1.5\%$, should the sampling plan be changed? If so, how much will be saved? How about if the process quality changes to $U_p = 0.5\%$ and $S_p = 0.5\%$?

4.3   Cost Parameters

This section describes each of the cost parameters required to select the action maximizing customer value. Section 4.4 does the same for the process parameters. The cost parameters are split into three groups: the release-discard decision parameters (option 1), the 100% inspection parameters (option 2), and the sampling inspection parameters (option 4). The release-discard parameters are:

N   =   Average lot size

P   =   Price unit sold for

D   =   Discard or salvage value of a unit

L   =   Loss occurred per defective unit sold

C   =   Per unit cost of manufacturing (fully loaded)

The first parameter, N, is the average lot size. It is not necessary that all lots be of the same size. However, for the calculations to be valid, the lot size should not be effected by changes in the process quality. A practice violating this assumption is to use smaller lots during periods of bad quality while using larger lots during periods of good quality.

The next two parameters, P and D, specify the income obtained by selling or disposing of units of products. Hopefully, most units are sold. P is price units are sold for. When prices vary depending on purchase quantity and so forth, P should be set to the average price. Some units might be discarded instead. If the units go in the trash, their discard value is zero. However, discarded units sometimes can be sold at a reduced price. The discard value D is the amount scrapped units can be sold for.

The next parameter, L, is the loss resulting from a customer receiving a single defective unit. This loss should include losses incurred by the company such as the refunding of the purchase price, the cost of providing replacement units, and the cost of processing returns and complaints. It should also include losses incurred by the customer such as the time and expense of returning the defective unit, and the downtime or loss business incurred until a replacement is received. The loss should also include intangible losses to the company such as loss of customer goodwill and loss of market share. Since the loss L includes the cost of refunding the purchase price or providing a replacement, it is always greater than the sale price P.

Typically, L is many times larger than the selling price P. Suppose one bought a thermos at a cost of $5. Upon getting home, one finds the interior is cracked. Losses incurred by the customer include 30 minutes travel time back to the store ($10), 20 miles put on the car ($5), and 15 minutes in the store getting a refund and buying another brand as a replacement ($5). Losses incurred by the store include time to process the refund ($2). Losses incurred by the company include the money refunded ($5) and the time for processing the return ($3). In addition, the customer tells 10 friends about the thermos, the line they had to wait in to get a refund, and the way they were treated by the store refund clerk. This results in a loss of sales to the store of $100 ($40 markup) and a loss of sales to the company of $5 ($3 markup). The total loss is $73, 15 times the price of the product.

Losses 10 and even 100 times the price are typical for defects that are not critical and cannot cause injury. For critical defects that can result in downtime or lost business or have the potential to cause injury, the loss can be a thousand, ten thousand, or even a million times the purchase cost. Take for example a $1 electrical component that has the potential of aborting a space launch if it fails.

It may be difficult to estimate L. It is generally hard to estimate the intangible losses such as the loss of customer goodwill and the loss of market share. In such cases, one can try a range of values for L to determine how L effects the action selected. For the television switch, Table 4.2 shows the effect of varying L. Regardless of whether L is $25, $50, $100, or $200, the best choice is to use a sampling plan. One can also determine how close to optimal the sampling plan n=80 and a=0 is for various values of L. Table 4.3 shows that the sampling plan n=80 and a=0 is within $20 of optimal for L between $25 and $100. This sampling plan is a good choice.

The next parameter, C, is the per unit cost of manufacturing. It should include all costs incurred in manufacturing the product including both direct costs such as material and labor costs and indirect costs such as overhead. The parameter C does not effect which action or sampling plan is best. It does, however, effect the values obtained for customer value. If a realistic value of C is used, customer value represents the gain from the manufacturing of the lot. This value can be compared to zero to determine if production should cease.

Table 4.2: Effect of L on Best Action

Loss (L)	Best Action	Customer Value
$10	Release All Lots	$2900.00
$25	n=43, a=0, 100% Inspect Rejects	$2805.88
$50	n=80, a=0, 100% Inspect Rejects	$2738.28
$100	n=139, a=0, 100% Inspect Rejects	$2645.96
$200	n=255, a=0, 100% Inspect Rejects	$2518.20

Table 4.3: Value of n=80 and a=0 For Various Values of L

Loss (L)	Customer Value n=80 and a=0	Customer Value Best Action
$10	$2828.24	$2900.00
$25	$2794.51	$2805.88
$50	$2738.28	$2738.28
$100	$2625.84	$2645.96
$200	$2400.94	$2518.20

The 100% inspection parameters are:

U_h = Per unit cost of the 100% inspection

S_h = Per lot setup cost of the 100% inspection

E = Efficiency of the 100% inspection (as percentage)

If units can be reworked, one final parameter is required:

R = Rework cost per unit

100% inspection is not possible if testing is destructive. When it can be performed, the 100% inspection can be characterized by three parameters: U_h, S_h, and E. The parameter U_h, is the per unit cost of performing the 100% inspection. It should include labor, material and equipment costs. S_h is the cost of preparing to 100% inspect a lot. If the 100% inspection can be performed on-line, S_h is typically zero. However, if performing the 100% inspection requires moving lots off-line, the cost of moving the material off-line, storing it, and transporting it to the inspection area are all part of S_h.

The parameter E is the percentage of defective units detected by the 100% inspection. Most 100% inspections are less than 100% efficient. It is not uncommon to have values of 80% or below for E.

When defective units can be reworked, R represents the average cost of reworking a unit. Rework is only cost effective if the cost of rework is less than the value gained, i.e., R < (P-D).

The sampling inspection parameters are:

U_s = Per unit cost of acceptance sampling

S_s = Per lot setup cost of acceptance sampling

The parameter U_s, is the per unit cost of selecting and testing units of product. It should include labor, material and equipment costs. S_s is the per lot setup costs. If acceptance sampling can be performed without altering the flow of production, then S_s is zero. If, however, the lot must be stored or delayed awaiting release, the cost of moving the material to storage, storage costs including carrying costs, and the cost of moving the material back from storage are all part of S_s.

4.4   Process Parameters

We have already seen how customer value depends on process quality. Figure 4.1 (page 61) shows a plot of customer value versus process quality for several different actions. To select the action maximizing customer value, one must determine how the process quality behaves.

The easiest way of characterizing process quality is by the process average. However, this is not sufficient when selecting sampling plans. If all lots are of the same quality, say p percent defective, then acceptance sampling is of no use. The best action is to either release all lots, discard all lots, or 100% inspect all lots. The best action can be determined by comparing p with the break even qualities. If, however, process quality varies, acceptance sampling can be of benefit by determining which lots are best released and which are best 100% inspected or discarded. The more the process quality varies, the larger the benefit provided by acceptance sampling. Selecting the best sampling plan or action requires estimating the variation in process quality.

The beta distribution is commonly used to describe the variation in process quality. The beta distribution can be characterized by two parameters: the average percent defective, U_p, and the standard deviation of the percent defective, S_p. Figure 4.5 shows the beta distribution for different combinations of these two parameters. Figure 4.2 (page 62) gives a third example.

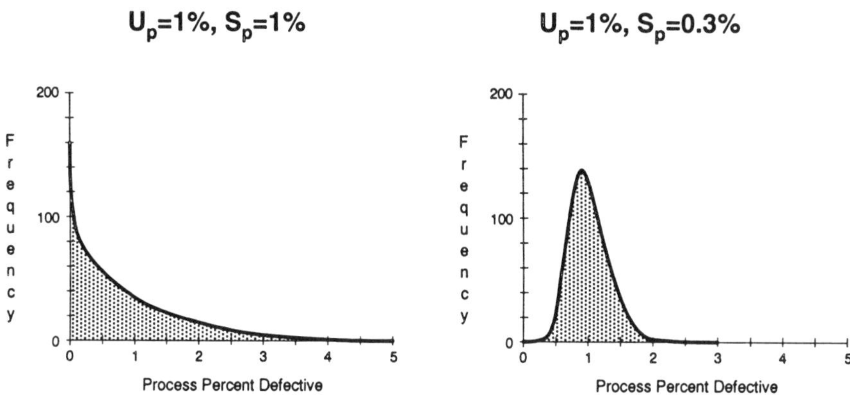

Figure 4.5: Two Different Beta Distributions for Process Quality

The program PROB can be used to calculate probabilities based on the beta distribution. These can be used to construct distribution plots. To calculate beta probabilities, select option 5 from the program's main menu. The program will then request p, U_p, and S_p. Based on these values, the program PROB calculates:

$\text{beta}(p \mid U_p, S_p)$ = relative frequency of p, i.e., the height of the curve on the plot of distribution function.

$\text{Beta}(p \mid U_p, S_p)$ = probability lot is less than or equal to p percent defective.

Screen 4.14 shows the program's results when p = 0.864%, U_p = 1%, and S_p = 2%. These values of U_p and S_p are from the television switch example. The value 0.864% is the break even quality between release and 100% inspect. Therefore, Beta(0.864% | 1%, 2%) = 0.724 is the probability that a lot is below the break even quality, i.e., is best released. The value of beta(0.864% | 1%, 2%) = 16.97 is the height of the curve at 0.864% in Figure 4.2 (page 62).

Screen 4.14: Calculation of Beta Probabilities (PROB)

```
*****************************************************************************
***                                                                      ***
***                 PROB Option 5 - Beta Distribution                    ***
***                                                                      ***
*****************************************************************************

   Both beta(p|Up,Sp) and Beta(p|Up,Sp) are calculated for the beta
   distribution.  It is assumed that the lot quality averages Up percent
   defective with a standard deviation of Sp percent defective.  Then:

      beta(p|Up,Sp)  =  height of beta curve at p percent defective
      Beta(p|Up,Sp)  =  probability lot is at most p percent defective


ENTER p   --> .864
ENTER Up  --> 1
ENTER Sp  --> 2


   Results:     beta(p|Up,Sp)   =   16.96830000
                Beta(p|Up,Sp)   =     .72383140



ENTER "r" TO REPEAT, "m" FOR MAIN MENU, OR "q" TO QUIT -->
```

Since p is limited to the range 0% through 100%, the standard deviation of p, S_p, is also limited in size. It must satisfy:

$$S_p \; < \; \sqrt{U_p\left(100 - U_p\right)}$$

If one has the results of past inspections, one can use these results to estimate both the average and standard deviation of the process percent defective. Suppose a sample of n units was selected from each of m lots. Let d_1, d_2, $\cdots$, d_m be the number of defective units found. The estimates of the average and standard deviation of the process percent defective are:

$$U_p \; = \; 100 \; \frac{\displaystyle\sum_{i=1}^{m} d_i}{m\,n}$$

$$S_p \; = \; 100 \; \sqrt{\frac{\displaystyle\sum_{i=1}^{m} d_i^2 \; - \; \sum_{i=1}^{m} d_i \; - \; \tfrac{1}{m}\left(1-\tfrac{1}{n}\right)\left(\sum_{i=1}^{m} d_i\right)^2}{m\,n\,(n-1)}}$$

$$\text{where:} \qquad \sum_{i=1}^{m} d_i \; = \; d_1 + d_2 + \cdots + d_m$$

$$\sum_{i=1}^{m} d_i^2 \; = \; d_1^2 + d_2^2 + \cdots + d_m^2$$

These formulas are from Lauer (1978). For the television switch example, samples of 50 units each were selected from 200 lots. The results are shown in Table 4.4. Base on these results:

$$n = 50 \qquad\qquad m = 200$$

$$\sum_{i=1}^{m} d_i \; = \; 101 \qquad\qquad \sum_{i=1}^{m} d_i^2 \; = \; 359$$

$$U_p = 1.01\% \qquad\qquad S_p = 2.06\%$$

Rounding gives $U_p = 1\%$ and $S_p = 2\%$. These were the values used in Sections 4.1 and 4.2.

Table 4.4: Inspection Results for Television Switch

Number of Defectives Found in Sample	0	1	2	3	4	5	6
Number of Lots	159	16	9	5	5	4	2

How accurate are these estimates? Formulas for the standard errors of these estimates are:

$$\mathrm{S.E.}\left(U_p\right) \; = \; 100 \sqrt{\frac{\left(\dfrac{U_p}{100}\right)\left(1-\dfrac{U_p}{100}\right) + (n-1)\left(\dfrac{S_p}{100}\right)^2}{m\,n}}$$

$$\mathrm{S.E.}\left(S_p\right) \; \approx \; 100 \sqrt{\frac{\left(1+\dfrac{\frac{U_p}{100}\left(1-\frac{U_p}{100}\right)}{(n-1)\left(\frac{S_p}{100}\right)^2}\right)\left(\left(\dfrac{S_p}{100}\right)^2 - \left(\dfrac{U_p}{100}\right)^2 + \dfrac{\left(1+2(n-1)\frac{U_p}{100}\right)^2}{4(n-1)n}\right)}{m}}$$

The true value for U_p will fall within ± 2 S.E.(U_p) of the estimate approximately 95% of the time. Likewise, the true value for S_p will generally fall within ± 2 S.E.(S_p) of its estimate. For the television switch, estimates of these standard errors are:

$$\mathrm{S.E.}\left(U_p\right) \; = \; 100 \sqrt{\frac{\left(\dfrac{1.01}{100}\right)\left(1-\dfrac{1.01}{100}\right) + (50-1)\left(\dfrac{2.06}{100}\right)^2}{200 \times 50}} \; = \; 0.175\%$$

$$\text{S.E.}\left(S_p\right) \approx 100 \sqrt{\frac{\left(1 + \frac{\frac{1.01}{100}\left(1 - \frac{1.01}{100}\right)}{(50-1)\left(\frac{2.06}{100}\right)^2}\right)\left(\left(\frac{2.06}{100}\right)^2 - \left(\frac{1.01}{100}\right)^2 + \frac{\left(1 + 2(50-1)\frac{1.01}{100}\right)^2}{4(50-1)50}\right)}{200}}$$

$$= 0.232\%$$

The true value of U_p is likely to fall in the interval (0.66%, 1.36%). The true value of S_p will generally be in the interval (1.48%, 2.64%). Even with a total of 10,000 samples, these estimates have a sizable error. A larger number of samples is preferable. Further, for such estimates to be accurate, the samples must be selected over a period of time long enough to experience the full range of process quality. Fortunately, it is not uncommon to accumulate such large amounts of data over extended periods of time.

Even when one has inspected a large number of lots, there is no guarantee that future lots will continue to behave in the same manner. No matter how many lots have been inspected, on some future lot a machine component might break or a bad lot of material might be used. This can result in a lot or series of lots uncharacteristic of those produced previously. The program VALUE allows a second distribution to be entered to account for these types of unusual but not impossible events. One can specify the average and standard deviation of the process quality under these uncharacteristic or problem conditions. One must also specify the percentage of the time the normal conditions apply, i.e., 95%, 99%, 99.9%, etc.

When the sample size varies, a different and more complicated set of formulas must be used. Suppose samples from m lots have been selected. From the i-th lot, d_i defectives where found in the sample of n_i units. Then the formulas for estimating U_p and S_p are as follows:

$$U_p = 100 \frac{\sum_{i=1}^{m}\left(\frac{d_i}{n_i}\right)}{m}$$

$$S_p = 100 \sqrt{\frac{\sum_{i=1}^{m}\left(\frac{d_i}{n_i}\right)^2 - \left(\frac{U_p}{100}\right)\sum_{i=1}^{m}\left(\frac{1}{n_i}\right) - \left(m - \sum_{i=1}^{m}\left(\frac{1}{n_i}\right)\right)\left(\frac{U_p}{100}\right)^2}{m - \sum_{i=1}^{m}\left(\frac{1}{n_i}\right)}}$$

where:

$$\sum_{i=1}^{m}\left(\frac{d_i}{n_i}\right) = \frac{d_1}{n_1} + \frac{d_2}{n_2} + \cdots + \frac{d_m}{n_m}$$

$$\sum_{i=1}^{m}\left(\frac{d_i}{n_i}\right)^2 = \left(\frac{d_1}{n_1}\right)^2 + \left(\frac{d_2}{n_2}\right)^2 + \cdots + \left(\frac{d_m}{n_m}\right)^2$$

$$\sum_{i=1}^{m}\left(\frac{1}{n_i}\right) = \frac{1}{n_1} + \frac{1}{n_2} + \cdots + \frac{1}{n_m}$$

When data is not available, one must choose reasonable values for U_p and S_p based on experience. Further one should explore the effect that errors or changes in these parameters have on the best action.

Exercises

(4.6)   Assume U_p = 2% and S_p = 3%. What is the probability that the process is 5% defective or less?

(4.7)   The results of inspecting 500 lots is shown in the table below. From each lot, 25 samples where selected.   Calculate estimates of U_p and S_p.

Number of Defectives	0	1	2	3	4	5	6	7	8	9	10
Number of Lots	402	58	24	8	2	0	0	2	3	0	1

(4.8)   Calculate the standard errors of the estimates in exercise 4.7. For both of these estimates, give intervals likely to contain the true value.

(4.9)   The results of inspecting 1000 lots is shown in the table below. For 814 of the lots, 13 samples were selected. For the remaining lots, 20 samples were selected.   Calculate estimates of U_p and S_p.

Number of Defectives	0	1	2	3	4	5	6	7	9
Number of Lots n=13	782	28	0	1	0	2	0	1	0
Number of Lots n=20	175	8	2	0	0	0	1	0	0

4.5   Formulas

This section may be skipped. It provides formulas for customer value and for the break even qualities. These formulas are specific to the cost model given in this book so they can not simply be referenced elsewhere. To those accustomed to such formulas, they help provide further understanding.

Customer value is:

$$\begin{matrix}\text{Customer} \\ \text{Value}\end{matrix} = \begin{matrix}\text{Income From} \\ \text{Sale of Units} \\ \text{and Discards}\end{matrix} - \begin{matrix}\text{Production} \\ \text{Costs}\end{matrix} - \begin{matrix}\text{Inspection} \\ \text{Costs}\end{matrix} - \begin{matrix}\text{Loss Due to} \\ \text{Sale of} \\ \text{Defectives}\end{matrix}$$

where:

$$\begin{matrix}\text{Income From} \\ \text{Sale of Units} \\ \text{and Discards}\end{matrix} = \left(\begin{matrix}\text{Number of} \\ \text{Units Sold}\end{matrix}\right) P + \left(\begin{matrix}\text{Number of} \\ \text{Units Discarded}\end{matrix}\right) D$$

$$\text{Production Costs} = N C + \left(\begin{matrix}\text{Number of} \\ \text{Reworked Units}\end{matrix}\right) R$$

$$\text{Inspection Costs} = \begin{matrix}\text{Cost of} \\ \text{Sampling Inspection}\end{matrix} + \begin{matrix}\text{Cost of} \\ \text{100\% Inspection}\end{matrix}$$

$$\begin{matrix}\text{Cost of} \\ \text{Sampling Inspection}\end{matrix} = \begin{cases} S_s + N U_s & \text{if sampling inspection} \\ 0 & \text{if no sampling inspection} \end{cases}$$

$$\begin{matrix}\text{Cost of} \\ \text{100\% Inspection}\end{matrix} = \begin{cases} S_h + N U_h & \text{if 100\% inspection} \\ 0 & \text{if no 100\% inspection} \end{cases}$$

$$\begin{matrix}\text{Loss Due to} \\ \text{Sale of} \\ \text{Defectives}\end{matrix} = \left(\begin{matrix}\text{Number of} \\ \text{Defective Units Sold}\end{matrix}\right) L$$

Customer value depends on the number of units sold, number of units discarded, number of units reworked, number of defective units sold, and on whether sampling and 100% inspections are performed. These in turn depend on the action taken and the process quality.

Suppose a lot that is p percent defective is released without any sort of inspection. Then:

Number of Units Sold	$=$	N
Number of Units Discarded	$=$	0
Number of Reworked Units	$=$	0
Number of Defective Units Sold	$=$	$N\,P/100$

No sampling or 100% inspection is performed

Therefore:

$$\text{Value}_{\text{release}}(p) \;=\; N\,P - N\,C - N\,\tfrac{P}{100}\,L$$

Similarly, customer value can be calculated for the other actions. Table 4.5 gives the customer value for the actions release, discard, 100% inspect discarding defectives, and 100% inspect reworking defectives. It is assumed that the lot is p percent defective. Then $N\,(1\text{-}P/100)$ is the number of good units in the lot and $N\,P/100$ is the number of defectives. The expression $N\,P/100\,(1\text{-}E/100)$ is the number of defective units missed by the 100% inspection and $N\,P/100\,E/100$ is the number of defectives found by the 100% inspection.

Table 4.5: Customer Value for Different Actions

Action	Income from sale of units and discards	Costs		Loss due to sale of defective units
		Production	**Inspection**	
Release lot	$N\,P$ -	$N\,C$ -	0 -	$N\,\tfrac{P}{100}\,L$
Discard lot	$N\,D$ -	$N\,C$ -	0 -	0
100% inspect, discard defectives	$\left(N - N\,\tfrac{P}{100}\,\tfrac{E}{100}\right)P + N\,\tfrac{P}{100}\,\tfrac{E}{100}\,D$ -	$N\,C$	$-\;\left(S_h + N\,U_h\right)$	$-\;N\,\tfrac{P}{100}\left(1 - E\!/\!100\right)L$
100% inspect, rework defectives	$N\,P$	$-\left(N\,C + N\,\tfrac{P}{100}\,\tfrac{E}{100}\,R\right)$	$-\;\left(S_h + N\,U_h\right)$	$-\;N\,\tfrac{P}{100}\left(1 - E\!/\!100\right)L$

For the television switch, Figure 4.1 (page 61) shows plots of customer value versus process quality for the first three actions. These plots were constructed using the formulas in Table 4.5. The plots of customer value are all straight lines. Of special interest are the points where these lines cross. They are the break even qualities.

The break even quality between releasing all lots and discarding all lots can be calculated as follows:

$$p_{release,discard} = 100 \frac{P-D}{L}$$

All break even qualities are expressed as percent defectives. This formula was obtained by setting the formulas for $Value_{release}(p)$ and $Value_{discard}(p)$ equal to each other and solving for p. A lot whose percent defective, p, is greater than $p_{release,discard}$ is best discarded. A lot better than $p_{release,discard}$ percent defective is best released. For the television switch the break even quality is:

$$p_{release,discard} = 100 \frac{5-0}{50} = 10\% \text{ defective}$$

When 100% inspection can be performed, one can rework or discard defectives. There is no break even quality between these two options. Instead, if rework is possible, rework should be performed only if (P-D) > R, i.e., the value gained by reworking is greater than the cost of reworking. When 100% inspection is possible, the break even qualities between releasing the lot and 100% inspection are:

$$p_{release,100\%-discard} = 100 \frac{U_h + S_h/N}{\frac{E}{100}(L+D-P)}$$

$$p_{release,100\%-rework} = 100 \frac{U_h + S_h/N}{\frac{E}{100}(L-R)}$$

If the lot percent defective is less than the break even quality, the best choice is to release the lot. If the process quality exceeds the break even quality, 100% inspection proves to be the better action. For the television switch, rework was not possible. The break even point between release and 100% inspection is:

$$p_{release,100\%-discard} = 100 \frac{0.25 + 100.00/1000}{\frac{90}{100}(50+0-5)} = 0.864\% \text{ defective}$$

The final pair of break even qualities are between 100% inspection and discarding the lot. They are:

$$p_{100\%-\text{discard,discard}} \;=\; 100\,\frac{P-D-U_h-\dfrac{S_h}{N}}{L-\dfrac{E}{100}(L-P+D)}$$

$$p_{100\%-\text{rework,discard}} \;=\; 100\,\frac{P-D-U_h-\dfrac{S_h}{N}}{L-\dfrac{E}{100}(L-R)}$$

If the process percent defective is less than the break even quality, the best choice is to 100% inspect the lot. If the process quality exceeds the break even quality, discarding the lot proves to be the best action. For the television switch, the break even point between 100% inspection and discarding the lot is:

$$p_{100\%-\text{discard,discard}} \;=\; 100\,\frac{5-0-0.25-\dfrac{100.00}{1000}}{50-\dfrac{90}{100}(50-5+0)} \;=\; 48.9\%\ \text{defective}$$

In addition to the actions in Table 4.5 (page 83), acceptance sampling can be performed. When performing acceptance sampling using the single sampling plan with sample size n and accept number a, the customer value is:

$$\text{Value}_{\text{sample}}(p) \;=\; \frac{\text{Value}_{\text{release}}(p)\ P_a(p)\ +\ \text{Value}_{\text{reject}}(p)\ (1-P_a(p))}{-\,(S_s+n\,U_s)}$$

Recall that $P_a(p)$ is the probability that the sampling plan accepts a lot that is p percent defective. The term $S_s + n\,U_s$ is the cost of sampling. Rejected lots may either be discarded, 100% inspected with defectives discarded, or 100% inspected with defectives reworked. Therefore the $\text{Value}_{\text{reject}}(p)$ can be either $\text{Value}_{\text{discard}}(p)$, $\text{Value}_{100\%\text{-discard}}(p)$, or $\text{Value}_{100\%\text{-rework}}(p)$.

Based on the above formulas, customer value can be calculated for any action as a function of the process quality p. These customer values are then averaged over the specified distribution of process quality in order to obtain the customer values reported using option 7 of VALUE.

4.6   Restricted Economic Sampling Plans

Sampling plans can be used to:

- Insure against the release of highly defective lots.

- Insure the average quality going to the customer is better than some specified level.

- Reject obviously bad lots.

- Improve the quality of the product going to the customer in order to maximize customer value.

Methods for selecting single sampling plans for the first three uses were covered in Chapter 3. This chapter covers the fourth use. However, it may also be desirable to choose sampling plans with more than one use in mind. This section shows how to select the sampling plan maximizing customer value under the restriction that its LTPD be less than or equal to a specified value. This sampling plan insures against the release of highly defective lots. Because of the restriction, it may not maximize customer value. However, in many cases, it comes very close.

Option 8 of the program VALUE can be used to select restricted economic single sampling plans. The sampling plan maximizing customer value for the TV switch was n=80 and a=0. This plan has an LTPD (Type-B OC curve) of 2.837. Suppose instead an LTPD of 1.5 or better is desired. Screen 4.15 shows the required input for option 8. All the cost and process parameters have already been entered.

Screen 4.16 shows the results. The sampling plan maximizing the customer value from those with an LTPD of 1.5% or better is n=153 and a=0. The corresponding customer value is $2719.45. This compares to a value of $2738.28 for n=80 and a=0. The cost of insuring that lots are all better than 1.5% defective costs $18.83 per lot.

Exercise

(4.10)   Assume one wants an LTPD of 2% or better for the TV switch. Find the single sampling plan with the desired LTPD that maximizes customer value. How does this sampling plan compare to n=80 and a=0 in terms of customer value?

Screen 4.15: Input for Restricted Economic Plan (VALUE)

```
******************************************************************************
***                                                                        ***
***       VALUE Option 8 - Select Most Economical Single Sampling Plan      ***
***                                                                        ***
******************************************************************************

   This option gives the most economical single sampling plan for the
   values entered.  If 100% inspection is possible, two sampling plans
   are selected: one when rejected lots are discarded and one when
   rejected lots are 100% inspected.  The best of these two plans is the
   one returned.  It is possible to restrict the search to only those
   sampling plans with a specified LTPD.

RESTRICT TO SPECIFIED LTPD?  ("y" or "n") --> y
ENTER LTPD AS PERCENT DEFECTIVE            --> 1.5
```

Screen 4.16 Results for Restricted Economic Plan (VALUE)

```
******************************************************************************
***                                                                        ***
***       VALUE Option 8 - Select Most Economical Single Sampling Plan      ***
***                                                                        ***
******************************************************************************

   When rejected lots are discarded, the most economical single
   sampling plan is:

      n    =        531
      a    =          4
      Value = $      1313.24      (per lot)

   When rejected lots are 100% inspected, the most economical single
   sampling plan is:

      n    =        153
      a    =          0
      Value = $      2719.45      (per lot)

   Rework is not possible so defective units should be discarded.

   The better plan is the one where rejected lots are 100% inspected.

PRESS ENTER KEY TO CONTINUE
```

4.7   Tallying Defects

The program VALUE assumes that defective units are tallied. What about the case of defects? There is a method that can be used to convert most problems into one that the program VALUE can handle. It will be demonstrated through an example.

Suppose one is inspecting airplane wings for defective rivets. The unit of inspection is a wing. Currently 1% of each wing is inspected, i.e., $n=0.01$. The number of defective rivets is tallied. A wing is rejected and sent back for rework if the number of defects is 2 or more.

The method is to redefine the unit of inspection to be a rivet. Since each wing contains approximately 5000 rivets, the above sampling plan is equivalent to inspecting 50 rivets and rejecting if there are two or more defective rivets. This means $n=50$ and $a=1$. Now defectives are tallied and the program VALUE can be used to select the action that maximizes customer value. Suppose that this action is to use the single sampling plan with $n_m=125$ and $a_m=2$. The subscript "m" signifies that this sampling plan is for the modified definition of a unit. Now one has to translate this sampling plan back.

To translate back requires the ratio of the original unit of production to the modified unit of production. This ratio will be denoted by r_n. For the airplane wing example:

$$r_n \quad = \quad \frac{1 \ \text{wing}}{5000 \ \text{rivits}} \quad = \quad 0.0002$$

Using r_n, the sample size n_o and accept number a_o for the original definition of a unit are:

$$n_o \quad = \quad n_m \times r_n \quad = \quad 125 \times 0.0002 \quad = \quad 0.025$$

$$a_o \quad = \quad a_m \quad = \quad 2$$

In the airplane wing example, a rivet was a natural modified unit of production. When the lot is continuous such as the inspection 0.5 cubic yards of cement, the modified unit of inspection could be 10 cubic inches, 1 cubic inch, or a cubic centimeter. The only requirement is that the chance of two or more defects occurring on a modified unit of production should be small.

4.8   Summary

Ideally all units sent to the customer should be defect free. However, most processes produce at least some defectives. One must decide whether to release, discard, 100% inspect, acceptance sample with rejected lots being discarded, or acceptance sample with rejected lots being 100% inspected. If 100% inspection is possible, one must also decide whether to discard or rework defective units. One way of making this decision is to select the action maximizing customer value.

Maximizing customer value requires determining the costs involved. The most critical cost is L, the loss due to the customer receiving a defective unit. L should include all losses incurred by the company, the customer, and any intermediates such as distributors. It should also include resulting losses in sales and market share. Otherwise, the companies short term profit is maximized instead of customer value. L is frequently difficult to determine. In which case a sensitivity analysis, varying L over a range, should be performed to determine the effect of L on the best choice.

Once the costs are known, the customer value can be plotted against process quality for each of the actions: release, discard, and 100% inspect. Each of these plots is a straight line. Of particular interest are the points where these lines cross. They are called the break even qualities. As indicated by these plots and the break even qualities, the best choice depends on how process quality varies.

The behavior of process quality can be described by U_p, the average percent defective, and S_p, the standard deviation of the process percent defective. A procedure is given for estimating these parameters from historical data. Obtaining good estimates requires a large amount of data (10,000 or more samples) over a period of time representative of the full range of variation in process quality. There is always the possibility that in the future something breaks or goes wrong producing product not typical of historical lots. Therefore, one can also specify the chance of such problems and how process quality might behave during such a problem.

Using this information, the program VALUE can be used to select the single sampling plans maximizing customer value when rejected lots are discarded and when rejected lots are 100% inspected. These sampling plans can then be compared to the actions release, discard, and 100% inspect. The best action is the one which maximizes customer value. If defective units can be reworked, they should be reworked only if R < (P-D).

Using the program VALUE, it is also possible to select single sampling plans maximizing customer value under the restriction that the LTPD be less than or equal to a specified value. Such sampling plans protect against the release of highly defective lots. They also frequently come close to maximizing customer value.

The program VALUE assumes that defectives are tallied. When defects are tallied instead, one must temporarily redefine the unit of production so that the chance of two defects on the same unit is small. One can then use the Program VALUE to select the sampling plan maximizing customer value. One must then translate the sampling plan back to the original unit of production using the procedure given.

References

Lauer, G. Nicholas (1978). "Acceptance Probabilities for Sampling Plans Where the Proportion Defective Has a Beta Distribution." Journal of Quality Technologies, Vol. 10, No. 2, pp. 52-55.

5

Practical Considerations

So far the evaluation and selection of single sampling plans has been covered. The next step is to learn how to use these sampling plans as part of a overall inspection program. With the background obtained so far, this chapter examines the broad issues of why one should acceptance sample and how to combine acceptance sampling with SPC and process monitoring. This chapter also covers the many practical considerations that must be addressed when applying acceptance sampling. These include where to place inspections, how to form lots, and how to select samples. Finally, this chapter highlights several common misuses of acceptance sampling and explains alternatives. The economic approach to selecting sampling plans from the previous chapter will prove enormously useful in exploring many of the topics in this chapter.

5.1   Why Acceptance Sample?

One reason to acceptance sample is that acceptance sampling can be the most economical action. Under certain conditions, acceptance sampling may prove more economical than releasing all lots or 100% inspecting all lots. For acceptance sampling to be the most economical action requires, among other things, that the process quality vary from lot to lot. The variation in process quality must be such that a sizable percentage of the lots fall on both sides of the break even quality.

Critics of acceptance sampling claim that all processes should be made stable and reduced to low levels of defects, i.e., made capable. For such processes there is no need for acceptance sampling. Ideally, all processes should be made stable and capable. However, it should also be recognized that processes exist which stretch existing technology. Such processes tend to be erratic and are often measured in terms of yields. Today it may be computer chips. Tomorrow it will be something else. But such processes will always exist and therefore there will always be a need for acceptance sampling.

Further, it takes time to achieve stable and capable processes. Achievement of a stable and capable process is a major achievement requiring extensive work. Ideally, all products and processes should be designed to be stable and capable at startup. In practice, most products and processes require some improvements and sometimes extensive improvements to make them stable and capable. New processes should have safeguards in place until such time as the process is proven stable and capable.

What about stable and capable processes or at least processes that are consistently below the break even quality? The best action is to release all lots. Should one really discontinue sampling and release all lots? That depends on how well past performance predicts future performance.

Consider the television switch example from Chapter 4. Suppose that the process has improved from before. Using the most recent 6 months worth of data, new estimates of U_p = 0.01% and S_p = 0.01% were obtained. This corresponds to 100 defects per million units produced. Based on these estimates, the action maximizing customer value is to release all lots. Should inspection really be discontinued? This depends on whether you believe future lots will also have U_p = 0.01% and S_p = 0.01%.

There is a certain amount of sampling variation in estimating U_p and S_p. However, much more important is the issue of how well past performance can be used to predict future performance. A past free of problems does not guarantee that no problems will occur in the future. Equipment wears and break, vendors go out of business, and so on. Even processes that have run for years without experiencing major problems, suddenly develop problems. What would happen if the television switch process started to experience problems even if only on a very infrequent basis? Suppose the process experiences problems an average of once every 100 lots and that when such a problem occurs, the lots averages 20% defective (U_p = 20%) with a standard deviation of 20% (S_p = 20%). Using the program VALUE with U_{p1} = 0.01%, S_{p1} = 0.01%, P_1 = 99%, U_{p2} = 20%, and S_{p2} = 20%, the action maximizing customer value changes to the sampling plan n=11 and a=0 where rejected lots are 100% inspected.

Table 5.1 compares the action release all lots with the sampling plan n=11 and a=0. When no problems develop (U_{p1} = 0.01%, S_{p1} = 0.01%, P_1 = 100%), the sampling plan n=11 and a=0 costs $5.89 more per lot than releasing all lots. If infrequent problems develop (U_{p1} = 0.01%, S_{p1} = 0.01%, P_1 = 99%, U_{p2} = 20%, S_{p2} = 20%), the sampling plan n=11 and a=0 saves an average of $74.65 per lot. This sampling plan results in a significant savings in the event of infrequent problems while only increasing costs slightly if no problems occur.

**Table 5.1: Comparison of Different Actions With and
Without the Occurrence of Problems**

Action	Customer Value U_{p1}=0.01%, S_{p1}=0.01%, P_1=100%	Customer Value U_{p1}=0.01%, S_{p1}=0.01%, P_1=99%, U_{p2}=20%, S_{p2}=20%	LTPD
Release all lots	$2995.00	$2895.05	-
n=11 and a=0	$2989.11	$2969.70	18.9%
n=57 and a=0	$2966.36	$2952.87	3.96%

The single sampling plan n=11 and a=0 maximizes customer value under one scenario. What about other scenarios? Since n=11 and a=0 has an LTPD of 18.9%, it will not provide much protection against 10% defective lots. When selecting sampling plans based on alternate scenarios, it is difficult to decide which scenario to use. The best sampling plan will change depending on the scenario selected. For this reason, rather than specifying different scenarios, it is generally better to specify an LTPD and select a restricted economic sampling plan.

For example, assume it is decided that protection is required against lots greater than 4% defective. Selecting the most economical sampling plan under the restriction that its LTPD be less than 4%, gives the sampling plan n=57 and a=0. This sampling plan is also listed in Table 5.1. It costs a little more under existing conditions, saves a little less money under the one alternate scenario, but protects against a wider range of alternate scenarios. This sampling plan appears to be a good choice under the circumstances.

This example points out that even when current conditions indicate inspection should be ceased, continuation of acceptance sampling may be advisable. When can one discontinue inspection? Based on these examples, it is not enough to have a good quality history. One must also evaluate the likelihood and magnitude of potential problems. Discontinuing inspection should require:

- A good quality history so that the action maximizing customer value under current conditions is to release all lots. This requires that lots be consistently below the break even quality.

- Controls be in place that ensure problems cannot occur or at least can not go undetected.

Suppose a bottle capping operation has an excellent quality history and discontinuation of inspection is being considered. One concern is removal force. Rigid controls on the dimensions of the bottle and cap in the forming operations protect against this problem. A second concern is missing caps. No control is currently in place that guarantees caps are not missing. Therefore, plans are made to install electric eyes on the line to detect this problem. As a result of these controls and the good history, inspection can be discontinued.

There is one last reason to acceptance sample. Suppose the two conditions given above have been meet. The process has a good quality history and the required controls are all in place. These controls may include the monitoring of process inputs such as temperature and pressure. They may also include monitoring of the outputs of the process. Frequently, control charts are used. Whenever information on the outputs is collected, even if for the purpose of monitoring or controlling the process, procedures should be in place to reject obviously bad lots of product. It would be hard to justify to a regulatory agency or in a court of law why a lot was released when data is on hand clearly indicating that the lot was bad.

Why acceptance sample? Several reasons are:

- Acceptance sampling may be the most economical choice.

- Acceptance sampling provides protection in the event that an unforeseen problem develops. This protection may be against the release of highly defective lots or a limit on the average outgoing quality. This protection is especially important during startup but can even be of benefit on processes averaging 100 defects per million or less.

- Acceptance sampling protects one from an unnecessary liability when samples are taken for other purposes such as process monitoring or control charting.

This is not to say that acceptance sampling should be the foundation of your quality program. Continuous improvement should be the foundation. Every effort should be made to prevent defects by eliminating the problems that produce defects. If more time and money is being spent on acceptance sampling than improvement activities or if management believes that acceptance sampling ensures the customers are getting good quality product, a dramatic change is needed. However, placing ones emphasis on improvement should not preclude acceptance sampling. Even world class companies continue to spend millions of dollars a year on acceptance sampling. The trick is to strike the proper balance.

5.2   Comparison of Methods for Selecting Plans

Table 5.2 shows the five methods for selecting single sampling plans and how they correspond to the different uses of acceptance sampling. To maximize customer value, one should select the most economical sampling plan. While there is one sampling plan maximizing customer value, there are a surprising number of sampling plans close to optimal. There is sizable latitude in selecting economical sampling plans. Table 5.3 shows a variety of sampling plans that are close to optimal for the television switch. The first sampling plan, n=80 and a=0, maximizes customer value. However, the other four sampling plans come close to optimal and are all significantly better than releasing all lots or 100% inspecting all lots. These sampling plans were selected to have AQLs significantly less than the break even quality of 0.864% and to have LTPDs significantly greater than the break even quality.

While interested in maximizing customer value, one may also desire to guard against highly defective lots. This is accomplished by specifying an LTPD and selecting the restricted most economical sampling plan. So long as the LTPD specified is significantly greater than the break even quality, this sampling plan will be close to optimal. In many cases, specifying the LTPD has little impact on the overall customer value. It can, however, result in large changes to the cost of inspection, losses due to defectives, and so on. One may find that the inspection costs double but are largely compensated for by a corresponding reduction in losses due to the release of defects.

Table 5.2: Methods of Selecting Sampling Plans

Methods of Selecting Sampling Plans	Uses of Sampling Plans			
	Insure against the release of highly defective lots	Insure average outgoing quality is better than specified value	Maximize customer value	Prevent liability due to release of obviously bad lots
Based on AQL and LTPD	X			
Based on AQL and AOQL		X		
Most Economical Sampling Plan			X	
Most Economical Sampling Plan With Specified LTPD	X		X	
Based on AQL and sample size				X

Table 5.3: Comparison of Different Sampling Plans for Television Switch Example

Sampling Plan	Value	AQL	LTPD	AOQL
n=80 and a=0	$2738.28	0.0641%	2.84%	0.457
n=125 and a=0	$2729.34	0.0411%	1.83%	0.293
n=32 and a=0	$2704.18	0.160%	6.94%	1.13
n=125 and a=1	$2723.96	0.285%	3.08%	0.670
n=200 and a=2	$2704.03	0.410%	2.64%	0.685
100% inspect all lots	$2555.00	-	-	-
release all lots	$2500.00	-	-	-

An alternate approach to selecting sampling plans that guard against highly defective lots is to specify the AQL and LTPD. So long as the AQL is significantly less than the break even quality, the sampling plan selected by this criteria should provide the same consumer protection and customer value as the matching restricted most economical sampling plan. If the LTPD is also significantly greater than the break even quality, this sampling plan will also be close to cost optimal. AQLs near or above the break even quality result in inefficient sampling plans.

5.3   Process Monitoring

A sizable percentage of the dollars spent on acceptance sampling are spent on sampling plans with minimal sample sizes. The most widely used sampling plan in industry is the single sampling plan n=13 and a=0. This sampling plan comes from Mil-Std-105E. This standard is the topic of Chapter 8. The sampling plan n=13 and a=0 results from trying to minimize the sample size given an AQL of 1%. How does one justify the use of such a sampling plan? Clearly it is not on the grounds of the protection provided. This sampling plan has an LTPD of 16.2%. It releases 5.2% defective lots half the time. It has an AOQL of 2.73%. This is not great protection.

We saw in Section 5.1 that such a sampling plan could maximize customer value. Such is the case when the process has a generally good history; however, there is still a chance of a major process failure. Such sampling plans are effective against a printer jamming for a period of time resulting in a series of misprinted cartons, or a cavity in a multi-cavity mold becoming dirty and resulting in a series of poorly molded parts. When major breakdowns or problems such as these are possible, some inspection should be routinely performed, no matter how good the historical data is. Before inspection can be discontinued entirely, safeguards and detection devices for all possible failure modes must be in operation.

Frequently such inspections are referred to as process monitoring. This highlights the fact that such sampling plans cannot insure each individual lot is acceptable. They can, however, detect major problems. In addition, the resulting data can be used to track the process quality over time in order to demonstrate improvement and to detect smaller sustained problems. Even when all major problems have been eliminated, it is still generally desirable to have some sort of routine inspection in place in order to track process performance.

Whenever samples are selected, no matter what the purpose, some sort of procedure should be in place to reject obviously bad lots.

Frequently, such data is control charted to detect trends or problems. An example is shown in Figure 5.1. The line at the top of the control chart is called an upper control limit. A point above the upper control limit indicates that the process has changed from its historical average. Control charts are covered in numerous texts. An outstanding reference is Wheeler (1986). Control charts exist for both counts of defectives and counts of defects.

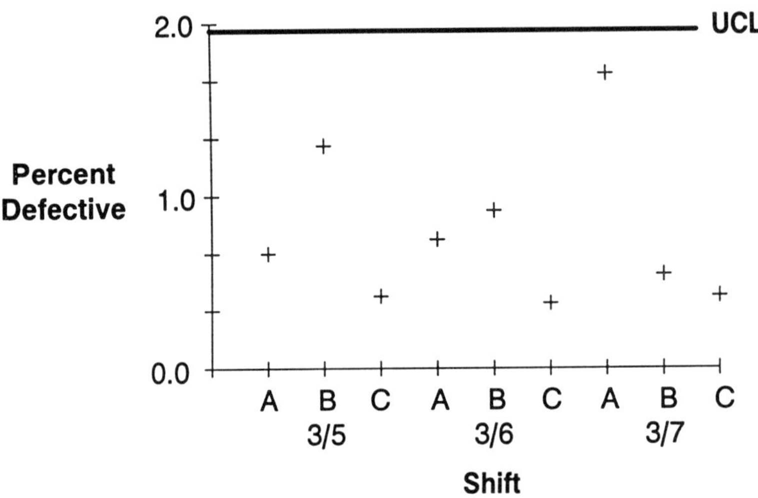

Figure 5.1: Example of a Control Chart

It is beneficial to use several control charts with different groupings of the data. Data can be grouped by lots, shifts, days, weeks, months, quarters, and years. Plotting small groupings such as lots or shifts results in faster detection times but limits the magnitude of the change that can be detected. Plotting large groupings such as months and quarters allows small changes to be detected, however, only on an infrequent basis and only if they are sustained. When the number of defects or defectives per group is frequently zero, CUSUM controls charts can speed the detection process. CUSUM charts are covered in Juran (1988).

When monitoring process quality, one should not curtail inspection of single sampling plans. Curtailing inspection makes estimating the process average difficult. When inspection is not curtailed, the following formulas are used to estimate the process average:

$$\text{Percent Defective} \;=\; 100 \times \frac{\text{Number of Defectives}}{\text{Number of Units Inspected}}$$

$$\text{Defects per Unit} \;=\; \frac{\text{Number of Defects}}{\text{Number of Units Inspected}}$$

The quality being estimated and monitored is the incoming quality. Following the inspection, the AOQ should be somewhat better. No foolproof procedure exists for estimating the AOQ except to pull new samples from released and reworked lots. One alternative is to estimate U_p and S_p using the procedure given in Section 4.4 (page 76) and then use option 7 of the program VALUE to estimate the AOQ. One must enter all the cost and process parameters along with the sampling plan before selecting option 7. Option 7 will then display an estimate of the AOQ. Only the sampling plan, the lot size, the efficiency of the 100% inspection, and whether units are reworked effect this estimate. It is incorrect to estimate the AOQ just using the samples from accepted lots. To see why, consider the sampling plan n=13 and a=0. All released lots have zero defects in their sample. Using only released lots to estimate the AOQ, the estimate of AOQ is always zero.

5.4   100% Inspections

100% inspections are often thought of as costly and time consuming. However, this is not always the case. Automatic 100% inspections, self checks and successive checks, sometimes using simple mistake proofing devices, may cost almost nothing on a per unit basis. Shingo (1986) gives an example of a device used to inspect the thickness of stem tighteners. This device consists of two bars placed above a chute at different heights. Tighteners catching on the first bar are too high. Tighteners failing to catch on the second bar are too short. The bars are at an angle so as to push the tighteners off the side of the chute into 3 different boxes.

Suppose the cost of installing such a pravice is $500. If 10 million units are inspected by the device, the per unit cost of the 100% inspection U_h is \$0.00005. The per lot setup cost S_h is zero. Further, suppose the device is 90% effective and that L = \$10, D = 0, and P = \$0.10. Then the break even quality, $p_{release,100\%\text{-discard}}$, is 0.000561% defective or 5.61 defects per million.

This example illustrates a key point:

> **Automatic or low-cost 100% inspections are cost effective even at low levels of defects and are always preferable to sampling inspection.**

Whenever possible, mistake proofing devices like the one given above should be used. They allow defect levels to be reduced beyond what can be economically achieved through acceptance sampling.

5.5   Defect Categories

It is common practice to group defects according to their severity. Mil-Std-105E classifies defects critical, major and minor. These categories are defined as follows:

Critical: Likely to result in hazardous or unsafe conditions or likely to prevent performance of a major end item such as a ship or aircraft

Major: Likely to result in failure or to reduce usability

Minor: Not likely to reduce usability

These correspond to safety, functional, and cosmetic defects. Defects of the same severity are grouped together and inspected as a group. Different sampling plans are selected for each group. Why divide defects into categories? Better protection is required for critical defects than for major and minor defects. The more severe the defect category, the smaller the LTPD or AOQL should be.

Using the cost model from Chapter 4, these categories correspond to significantly different values of L. L is the cost of releasing a defective unit. Critical defects are life and health threatening. The parameter L can be hundreds of thousands of dollars if a value can be put on it at all. Major defects are functional defects. L is likely to be several to 100 times the price of the product. Minor defects are cosmetic in nature. L is likely to be from a fraction to several times the price of a unit. The most economical sampling plan will be different for these different categories.

5.6   Where to Place Inspections

Acceptance sampling can be used for receiving inspections, in-process inspections, and final inspections. If the process consists of multiple step, acceptance sampling can be performed following each step. With all these options, where does acceptance sampling do the most good? If only a single defect type is being inspected for, the rule is to inspect immediately after the last process step effecting the defect type. Delaying inspection any further only allows additional value to be added to defective units before they are rejected.

If a defect is only related to the materials used, it should be inspected for at receiving, or even better, at the suppliers plant. For example, consider a process for manufacturing pencils using graphite leads, wood slates, and eraser assemblies. The process steps are to cut the graphite leads to the correct length, form the wood slats, glue the slats together with the graphite between them, sand, paint, and add an eraser. The hardness of the lead should be inspected at receiving or accepted by certification from the supplier. None of the subsequent process steps effect the hardness. Care should be taken to ensure the testing is consistent with in-house testing.

If a defect type is effected by the process, acceptance sampling is most economical if performed immediately following the last process step effecting the defect type. The overall length of the pencil can only be inspected for at the end of the process, i.e., final inspection. This is because the final processing step, adding the eraser assembly, effects pencil length. In contrast, break force should be inspected immediately following the gluing operation. Subsequent steps, including sanding and painting, do not effect break force.

If several defect types are inspected for, it may be simpler to group them together and do all the testing at one time. A trade off must be made between the value added to defective units and the reduction in inspection costs. In special cases it may be necessary to delay an inspection due to measurement requirements. The product may require a certain amount of time to cure before testing can be performed. Sometimes subsequent processing is required before measurements can be taken. For example, in making sandwich bags, the seals are tested using a burst test. Following side seal formation, the bottom seals must be formed before testing can be performed.

5.7   Forming Lots

When sampling to insure against the release of highly defective lots or to insure the AOQ is below some specified value, the protection provided by a sampling plan is not greatly effected by the lot size, especially for larger lots. As a result, one way of reducing inspection costs is to increase the lot size. There is no need to change the sampling plan as the lot size increases. The protection remains the same. As a result, doubling the lot size, cuts in half the amount of inspection performed.

When selecting economical sampling plans, things are not as simple. Keeping the sampling plan the same reduces the cost of acceptance sampling. However, as lot size increases, the cost of 100% inspecting lots, the cost of discarding lots, and the cost of releasing defectives all increase proportionally. It is beneficial to use some of the reduction in acceptance sampling costs to make better accept/reject decisions so as to reduce these other costs as well.

Table 5.4 shows the effect of increasing lot size on the most economical sampling plan for the television switch. Customer value is given on a per unit basis so the results are comparable. The benefit of a ten-fold increase in lot size is a savings of $69 per lot of 1,000 units. Less than a quarter of this savings is obtained by increasing the lot size while keeping the sampling plan the same.

Table 5.4:   Effect of Lot Size on Most Economical Sampling Plan for the Television Switch. Assumes Process Quality is Unaffected by the Lot Size Change.

Lot Size	Most Economical Sampling Plan	Value per Unit of Most Economical Sampling Plan	AQL	LTPD
1,000	n=80 and a=0	$2.738	0.0641%	2.84%
3,000	n=127 and a=0	$2.783	0.122%	1.33%
10,000	n=292 and a=1	$2.807	0.122%	1.33%

If increasing lot size decreases costs, what incentive is there to keep the lot size small? With large lots, short periods of bad product are easily missed as they are combined with other lots. The result is more defects going to the customer. If the group of bad product is detected, the entire lot must now be 100% inspected, increasing the inspection costs. The cost of releasing defectives and the cost of 100% inspecting rejected lots can increase dramatically if the new larger lots are not homogeneous. These costs can easily over shadow the reduction in acceptance sampling costs.

In doing the above analysis, it was assumed that the distribution of process quality was not effected by the change in lot size, i.e., the new larger lots are homogenous. This represents the scenario most favorable to increasing the lot size. But what if they are not homogenous? The worse scenario is that the quality of adjacent lots are entirely independent of each other. In this case, the standard deviation of process quality for lots m times the original size is

$$\text{Standard of Lot Quality for Large Lot} \;=\; S_p \big/ \sqrt{m}$$

where S_p is the standard deviation of process quality for the original lot size. The standard deviation of the combined lot is smaller since the individual lots making up the combined lot average each other out. Assuming independent lots, Table 5.5 shows how increasing the lot size effects the most economical sampling plan. The result of increasing the lot size is to decrease customer value by $50 per lot of 1,000 units.

Table 5.5:   Effect of Lot Size on Most Economical Sampling Plan for the Television Switch. Assumes Adjacent Lots are Independent of Each Other.

Lot Size	Process Quality	Most Economical Sampling Plan	Value per Unit of Most Economical Sampling Plan
1000	$U_p = 1$, $S_p = 2$	n=80 and a=0	$2.738
3,000	$U_p = 1$, $S_p = 1.15$	n=161 and a=0	$2.705
10,000	$U_p = 1$, $S_p = 0.632$	n=315 and a=0	$2.688

Most cases fall somewhere in between these two scenarios. Adjacent lots tend to be similar in quality to each other but are not identical. Whether the lot size should be increased depends on how rapidly the quality changes across a series of lots. The best lot size is the largest lot size still yielding relatively homogeneous lots.

5.8   Selecting Samples

To ensure the OC curves, ASN curves, and AOQ curves are valid, care must be taken in the selection of the samples to ensure they are representative of the lot as a whole.   One way to ensure representative samples is to select random samples.

A random sample requires that each unit have the same chance of being selected and that the selection of a particular unit has equal effects on the chances of the others units being selected. Suppose one is told to select five samples from the beginning of the lot and five samples from the end. This is not a random sample. Units towards the middle of the lot are not as likely to be selected as units near the ends of the lot. Suppose one is told to select four cartons and then pull five units from each carton. Neither is this a random sample. If one unit from a carton is selected, other units in the same carton are more likely to be selected while those in other cartons are less likely to be selected.

One can use random number generators or tables of random numbers to select random samples. One procedure is to number the units 1 to N where N is the lot size. Then generate random numbers from 1 to N to determine which samples to include. Replace repeated values with new random numbers. Generally such a formal procedure is not used. Instead it is left to the judgment of the inspector to ensure that the sample is "random". Certain difficulties can arise in doing this. First, not all units may be equally accessible. In this case the tendency is to select the more accessible units such as those on the top of a carton or drum. It is generally better to select the samples from the line before they are packaged. Second, certain defects may be visible to the inspector as they are selecting the sample. The inspector must be blinded to such information.

Continuous lots such as a drum of powder represent a special problem. One way of sampling bulk product is to take periodic samples as the drum is filled or emptied. The other approach is to lower a special sampler into the drum to sample small quantities until the required sample size is obtained. Quantities should be selected from the top, bottom, sides, and middle of the drum to ensure a representative sample. If the tank has been thoroughly mixed so that it contains a homogeneous blend, a single sample of the required size can be taken instead.

An alternative to random sampling is stratified sampling. When pulling a stratified sample, the lot is broken into subgroups of equal size and the same number of samples are selected from each subgroup. For example, suppose a lot takes 8 hours to produce and that a sample of 32 units is required. Then every hour the inspector selects a sample of four units. Stratified sampling requires that within each subgroup the samples be selected randomly. Stratified sampling ensure the samples are more evenly spread out across the lot. It provides a better estimate of the lot quality. It can be used as a replacement for random sampling and in many cases is preferable to random sampling. Lots can be stratified by time, cavity, nozzle and so on.

A common practice is to take periodic samples such as every 50th unit. Periodic samples are similar to stratified samples in that they ensure the samples are more evenly spread out across the lot. They differ from stratified samples in that the samples within a subgroup are not random. Care should be taken when using periodic samples. Suppose every 50th tablet is selected from a tabletting machine. If the tabletting drum has 25 cavities, all the samples are from the same cavity. The period between samples should be selected to ensure equal coverage of all cavities. Another pitfall is the fact that it is known which units are to be inspected when and even before they are made. This can result in special attention being given to these units so that they are not representative of the rest of production. Periodic samples offer a special advantage in that the selection of the samples can be easily automated.

A common practice for sampling product that is already packed is to specify that a certain number of cartons be sampled and that multiple samples be selected from each carton. Defects that are randomly dispersed throughout the cartons do not present a problem. However, the presence of defects that cluster together in groups or occur in runs can lead to dramatic reductions in the level of protection. Suppose the sampling plan n=20 and a=0 is used. This sampling plan has an LTPD of 10.9% defective. Lots that are 10% defective are accepted about 10% of the time. The standard procedure is to select 4 cartons and pull 5 samples from each carton. Suppose a lot consists of 10 cartons, one of which contains all defectives and the other 9 of which are free of defects. This lot is 10% defective and should have a 10% chance of acceptance. However, using the standard procedure, the lot is accepted if none of the four cartons selected is the defective carton. The probability of accepting this lot is 60%. In this worse case, the protection provided is that of the single sampling plan n=4 and a=0.

5.9   Samples Sizes for Special Studies

Many times special studies are run on new processes to demonstrate they are capable of producing good product. Typically one wants to make a statement such as with 90% confidence, the process can produce product that is less than 1% defective. The sampling plan tables and the program SINGLE can be used to determine the number of samples required and the criterion that must be passed in order to make such a statement.

Suppose one wants to make the statement that with 90% confidence the process can produce product that is less than P% defective. One should then select a sampling plan with an LTPD of P% defective. Table 3.3 (page 47) gives several sampling plans with LTPDs close to 1% including 200/(0,1), 315/(1,2), and 500/(2,3). Selecting the required number of samples and passing the acceptance criterion allows one to make the desired statement. These sampling plans differ with respect to their AQLs. One should select an AQL that gives the process a reasonable chance of passing. If one believes the process is averaging several hundred defects per million, one could use the minimum sample size of 200. If one believes the process is averaging closer to 0.1% defective, the sampling plan with 315 samples having an AQL of 0.113% defective should be used instead.

To obtain confidence statements with other than 90% confidence, the program SINGLE must be used. Suppose one wants to make the statement that with C% confidence the process can produce product that is less than P2% defective. Further suppose that one wants at least a probability P of accepting the process if it is running P1% defective or better. Then start the program SINGLE as before and select option 6. Specify that defectives are tallied and select the Type-B OC curve. Next enter P1 for the AQL and P2 for the LTPD. Also specify that nonstandard definitions of the AQL and LTPD are to be used. Last, enter P for the probability of acceptance at the AQL and (1-C/100) as the probability of acceptance at the LTPD. Passing the selected sampling plan allows one to make the desired statement.

Suppose one wants to state that with 99% confidence that the process can produce product that has fewer than 1000 defects per million, i.e., is less than 0.1% defective. Further, suppose one believes the process is running around 100 defects per million (0.01% defective) and that a probability of acceptance of 90% is desired at this point. Screen 5.1 shows the input required by SINGLE. The resulting sampling plan is n=8403 and a=2. If one selects 8403 samples and finds 2 or fewer defectives, one can state that with 99% confidence, that the process can produce less than 1000 defects per million.

Screen 5.1: Selecting Sample Size for Special Study (SINGLE)

```
*****************************************************************************
***                                                                     ***
***        SINGLE Option 6 - Select Sampling Plan Based on AQL and LTPD  ***
***                                                                     ***
*****************************************************************************

    This option selects the single sampling plan minimizing the ASN curve
    from among those satisfying the following set of conditions:

        Actual AQL is greater than or equal to the specified AQL
        Actual LTPD is less than or equal to the specified LTPD

    One must specify whether to use the Type-A or Type-B OC curve for
    defects or defectives.  The standard definitions of AQL (95% chance of
    acceptance) and LTPD (10% chance of acceptance) are used unless indicated
    otherwise.

TALLY? (1=defectives,2=defects)              --> 1
ENTER TYPE OF OC CURVE. ("A" or "B")         --> b
ENTER AQL   AS PERCENT DEFECTIVE             --> .01
ENTER LTPD AS PERCENT DEFECTIVE              --> .1
USE .95 & .1 FOR AQL-LTPD? ("y" or "n") --> n
ENTER PROBABILITY ACCEPT AT AQL              --> .90
ENTER PROBABILITY ACCEPT AT LTPD             --> .01
```

Table 5.6 shows the sample sizes and acceptance criteria for a variety of confidence levels and values of P2. Special startup studies are a good practice. It is during the initial couple of lots that problems are most likely to occur. It pays to allocate extra inspection resources to these lots. Once the process has proven itself, inspection can be reduced to a more economical level.

There are also times when one wants to make such confidence statements based on historical data. Suppose one has accumulated 10,000 samples of which 13 were defective. What can one say about the process? The program SINGLE can be used to make a confidence statement. Start the program as before and select option 1 to enter a sampling plan. Specify that defectives are tallied and enter 10,000 for the sample size. Enter 13 for the accept number. Next select option 2 and specify a Type-B OC curve. Screen 5.2 shows the result. Based on the values shown, one can state that with 90% confidence that the process is averaging less than 0.190% defective. One can also state with 95% confidence that the process is averaging less than 0.207% defective. Confidence statements and sample sizes can be determined for defect rates in a similar fashion.

Table 5.6: Sample Sizes and Accept Numbers for Confidence Statements

Upper Limit (P2)	Confidence Level								
	90%			95%			99%		
	n	a	AQL	n	a	AQL	n	a	AQL
1%	230	0	0.0223%	300	0	0.0171%	460	0	0.0111%
	390	1	0.0912%	470	1	0.0757%	660	1	0.0539%
	530	2	0.154%	630	2	0.130%	840	2	0.0974%
	2500	18	0.498%	3100	22	0.508%	4700	31	0.496%
0.5%	460	0	0.0111%	600	0	0.0855%	920	0	0.00558%
	780	1	0.0456%	950	1	0.0374%	1300	1	0.0273%
	1060	2	0.0772%	1300	2	0.0629%	1700	2	0.0481%
	5000	18	0.249%	6300	22	0.250%	9300	31	0.251%
1000 dpm	2300	0	0.00223%	3000	0	0.00171%	4600	0	0.00112%
	3900	1	0.00911%	4700	1	0.00756%	6600	1	0.00539%
	5300	2	0.0154%	6300	2	0.0130%	8400	2	0.00974%
	25000	18	0.0498%	31000	22	0.0507%	47000	31	0.0496%
500 dpm	4600	0	0.00112%	6000	0	0.000855%	9200	0	0.000557%
	7800	1	0.00456%	9500	1	0.00374%	13000	1	0.00273%
	10600	2	0.00772%	13000	2	0.00629%	17000	2	0.00481%
	50000	18	0.0249%	63000	22	0.0250%	93000	31	0.0251%
100 dpm	23000	0	0.000224%	30000	0	0.000170%	46000	0	0.000110%
	39000	1	0.000909%	47000	1	0.000754%	66000	1	0.000539%
	53000	2	0.00154%	63000	2	0.00130%	84000	2	0.000975%
	250000	18	0.00498%	310000	22	0.00507%	470000	31	0.00496%

Screen 5.2: Determining Confidence Statements (SINGLE)

```
                         SUMMARY STATISTICS
                  (Type-B OC Curve for Defectives)

     Current Sampling Plan:  n =        10000,    a =      13
                             tally defectives


                                             Process
              Percentile    Symbol     Percent Defective


                 95.0%      AQL               .0846578
                 90.0%                        .0947126
                 50.0%      IQ                .1366764
                 10.0%      LTPD              .1895216
                  5.0%                        .2065959


     AOQL =          .0867065%   (Assumes 100% inspection of rejected
                                  lots with all defectives being found
                                  and repaired plus a large lot size.)

PRESS ENTER KEY TO CONTINUE
```

5.10   Effect of Inspection Errors

When inspecting one can make two types of errors: classify a defective unit as good and classify a good unit as defective. Let p_1 and p_2 denote the probability of making these two errors where:

p_1 = probability of classifying good unit as defective

p_2 = probability of classifying defective unit as good

Further, let p denote the true process percent defective. Then the observed percentage of units classified as defective, denoted p^* is:

$$p^* = p_1 \times (100 - p) + (1-p_2) \times p$$

In this formula, 100-p is the percentage of good units which is multiplied by the probability of classifying these as defective. The term p is the percentage of bad units which is multiplied by the probability of classifying these as defectives, i.e., $1-p_2$. Adding these together gives the total number of units classified as defective.

The OC curves, ASN curves, AQLs, and LTPDs we have been using all assume that no inspection errors occur, i.e., p_1=0 and p_2=0. In this case p^*=p. When inspection error does occur, p^* and p are no longer the same. The OC curve, ASN curve, AQL, and LTPD are all in terms of the observed percent defective p^*. To translate these in terms of the true percent defective, one first needs to solve for p:

$$p = \frac{p^* - 100 \times p_1}{1 - p_1 - p_2}$$

Using this formula, one can calculate the actual AQL and LTPD:

$$AQL_{actual} = \frac{AQL_{calulated} - 100 \times p_1}{1 - p_1 - p_2}$$

$$LTPD_{actual} = \frac{LTPD_{calulated} - 100 \times p_1}{1 - p_1 - p_2}$$

For example, suppose that one is using the single sampling plan n=50 and a=1. Assuming no inspection errors, the AQL of this sampling plan is 0.715% and the LTPD is 7.56%. If 0.5% of all good units are falsely classified as defectives (p_1=0.005) and 10% of the defective units are missed (p_2=0.1), then:

$$\text{AQL}_{\text{actual}} \quad = \quad \frac{0.715 - 100 \times 0.005}{1 - 0.005 - 0.1} \quad = \quad 0.240\% \text{ defective}$$

$$\text{LTPD}_{\text{actual}} \quad = \quad \frac{7.56 - 100 \times 0.005}{1 - 0.005 - 0.1} \quad = \quad 7.89\% \text{ defective}$$

To draw OC curves and ASN curves based of the true percent defective, one must relabel the bottom axis in terms of the true percent defective.   When the true percent defective (p) is 0% defective, $p^* = p_1 \times 100 = 0.5\%$. When the true percent defective (p) is 100% percent defective, $p^* = (1-p_2) \times 100 = 90\%$. When the observed percent defective (p^*) is 10%, the true percent defective is:

$$\frac{10 - 100 \times 0.005}{1 - 0.005 - 0.1} \quad = \quad 10.6\% \text{ defective}$$

Figure 5.2 shows the OC curve for n=50 and a=1 in terms of the true percent defective. Figure 2.5 (page 12) showed the OC curve in terms of the observed percent defective. The only difference is the labeling of the bottom axis.

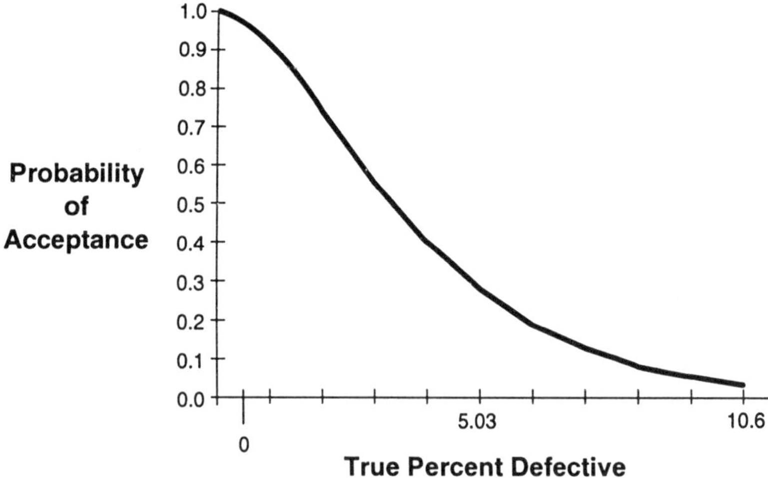

**Figure 5.2: Type-B OC Curve of Sampling Plan n=50 and a=1
With Inspection Errors (p_1=0.005, p_2=0.1)**

5.11   Incentive for Improvement

The purpose of acceptance sampling is to improve the quality of the lots going to the customer. Acceptance sampling does this directly by rejecting some lots that must be either discarded or 100% inspected. The end result is that the quality of the lots going to the customer is better than that coming into the sampling plan.

Acceptance sampling can also improve quality indirectly by providing incentives for improvement. In this case it is the quality of the lots coming into the sampling plan that improves. Using acceptance sampling to provide incentive for improvement is explored in this section. The other side of the coin is that as improvement takes place, this improvement should be rewarded. Certain sampling practices can act as barriers to improvement and sometimes even penalize improvement. Those practices are identified so that they can be avoided.

Improving quality results in fewer lots being rejected and either discarded or 100% inspected. Scrap rates and inspection costs are decreased. Acceptance sampling provides incentive to improve and maintain the process average below the AQL of the sampling plan. A sampling plan with an AQL of 1% defective and LTPD of 5% defective puts pressure on the producer to maintain the process average below 1% defective and insures individual lots are less than 5% defective. The scrap and inspection costs only provide an incentive if the producer bears these costs. For incoming goods, rejected lots should be charged to the supplier or returned to the supplier for 100% inspection. For in-process and final goods, the same is true for the production department responsible for the rejection of the lot.

Acceptance sampling can be used to translate the cost of releasing defectives including complaints, returns, loss sales and so on into costs that are more tangible to production such as scrap and labor hours. It is important that production recognize this so that the sampling plan is not viewed as an obstacle erected by the Quality department. The sampling plan should be viewed as a reflection of the customer's needs and desires. Acceptance sampling should not be used to create an artificial incentive.

The incentive provided by the AQL is frequently abused. One abuse is to try and drive improvement efforts by continually tightening the AQLs. This practice is harmful in many ways. First, the sampling plan losses credibility and becomes viewed as a tool used by the

Quality department to control production. Second, as improvements are made, this practice fails to reward them. As the process average improves, the scrap will be reduced. However, then changing to a tighter AQL, increases the scrap back to where it was. Further, continuing to tighten the AQL will increase the inspection costs. The result is that more inspection effort is spent on higher quality processes than on poorer quality processes where the inspection could be of greater benefit. Finally, the resulting combative crisis driven environment does not support structured problem solving involving cross functional teams.

Any inspection program should keep the following points in mind:

- The bulk of the inspection effort should be concentrated on the more problematic processes and new processes where it will do the most good.

- As process quality improves, the scrap, the amount of inspection and the frequency of inspection should be reduced to reward the improvement effort.

Acceptance sampling can also act as a barrier to improvement by serving as a crutch to lean against. Many find solace in the belief that no matter what else happens, the sampling plan will make everything all right. This results in shoddy practices such as allowing questionable periods of production to continue down the line to see if they pass the sampling plan rather than isolating them for special treatment. For those who use acceptance sampling as a crutch, there can be great distress in seeing the actual OC curves of the sampling plans and in understanding their limitations. Removing the acceptance sampling crutch starts the process of improvement by raising questions such as:

What is the quality of the product going to the customer?	$\Rightarrow$	Generally, only slightly better than that produced.
How does one insure the customer only receives good product?	$\Rightarrow$	Make only good product.
How does one make only good product?	$\Rightarrow$	Eliminate the problems that lead to defects.

5.12   Zero Acceptance Number Sampling Plans

With the emphasis on achieving zero defects or defect-free product, zero acceptance number sampling plans have been receiving increasing attention. It appears to be inconsistent to strive towards zero defects and yet accept lots whose samples contain defects. However, there is a fundamental difference between zero defects and a zero acceptance number.

Zero defects is an ideal to aim for in producing the product. Zero defects can be achieved if the problems leading to defects are eliminated. However, once the product has been produced, it must be dispositioned. The sampling plan can not guarantee zero defects. It does not have control over the production of defects or the disposition of individual defects. The only options open to the sampling plan are to release, 100% inspect, or discard the entire lot.

Much time and effort has been invested in the lot. What one is willing to accept of an already produced lot is logically different from what one desires of new lots. What one is willing to accept should be based on economics with the goal of providing the best value to the customer. The customer losses and inconvenience must be an integral part of this decision process. One is not doing the customer a favor by using a zero acceptance number sampling plan in the name of zero defects if the resulting improvement in quality costs the customer ten times what it is worth.

Squeglia (1986) gives a tables of zero acceptance number sampling plans that can be used as replacements for the Mil-Std-105E plans. The zero acceptance number replacements have the same LTPD as the MIL-Std-105E plans they replace. They also have smaller sample sizes. For example, the replacement for the Mil-Std-105E sampling plan n=125 and a=3 is the zero acceptance number plan n=42 and a=0. Both have LTPDs of 5.3% defective. Both plans provide the same protection against the release of highly defective lots. However, they differ in their AQLs. The Mil-Std-105E plan has an AQL of 1.10% defective while the zero acceptance number plan has an AQL of 0.122% defective.

Zero acceptance number sampling plans are described by Squeglia as being "originally designed and used to provide equal or greater consumer protection with less inspection than the corresponding Mil-Std-105D sampling plans." The replacement sampling plan in the above example does just this. However, because of the sizable reduction in the AQL, use of this replacement plan can significantly increase the number of lots rejected. What is not considered by this approach is the increased cost of 100% inspecting rejected lots or the increased cost of discards. Simply reducing sampling inspection costs can result in poor value to the customer.

Before using a zero acceptance number sampling plan one should compare the new AQL to the process average. If the process average is significantly below the new AQL, the zero acceptance sampling plan is probably the most economical sampling plan for the specified LTPD, or at least close to it. As one reduces the levels of defects for a process, the most economical sampling plan will eventually be a zero acceptance number sampling plan.

Zero acceptance number sampling plans should be the reward for achieving improvement. They are not the tool for effecting such improvement. Going to zero acceptance number sampling plans to help drive the process to zero defects is akin to trying to drive down defect levels by lowering AQLs. The ill effects of this policy were covered in the previous section.

When using the program VALUE, zero acceptance number plans will be selected as a matter of course when they are most economical. Such an example can be found in Section 5.1 (page 91). Zero acceptance number plans can also be found in the first column of Table 3.3 (page 47). Further, they can be selected for a specified LTPD by using option 6 of the program SINGLE. A value close to zero should be entered for the AQL (e.g., 0.000001).

5.13 Summary

Acceptance sampling is used because:

- Acceptance sampling may be the most economical choice.

- Acceptance sampling provides protection in the event of unforeseen problems.

- Acceptance sampling protects one from an unnecessary liability when samples are taken for other purposes such as process monitoring or control.

One requirement for discontinuing acceptance sampling is that there should be a good quality history so that releasing all lots maximizes the customer value. Good historical performance does not, however, guarantee problems will not develop down the road. Therefore, a second requirement is that controls should be in place to ensure problems cannot occur or at least go undetected.

While there is one single sampling plan that maximizes customer value, many sampling plans are close to optimal. This provides some latitude in determining which sampling plan to use. Sampling plans whose LTPDs are above the break even quality and whose AQLs are below the break even quality are generally close to optimal. Restricted economic plans will be close to optimal so long as the specified LTPD is above the break even quality.

Process monitoring cannot be justified based on its lot by lot protection. It can however, detect major problems and is required in order to track process performance. Use of the single sampling plan n=13 and a=0 is in essence process monitoring.

When the cost of a 100% inspection can be reduced close to zero, 100% inspection is the most economical approach even for a process with only a small number of defects per million. Whenever possible, such low cost mistake proofing and automatic 100% inspections should be implemented.

Defects are frequently classified as critical, major or minor depending on the severity of their effect. The cost of releasing a defect to the customer L is different for defects of different severity. Therefore, different sampling plans will maximize customer value.

The inspection for a defect should be performed immediately after the last manufacturing step effecting the defect. For raw materials, the inspection should take place at the suppliers location. Sometimes, when inspecting multiple defect types, it is convenient to group the inspections into a single test station.

Increasing the lot size frequently effects the distribution of lot quality and can result in more units being 100% inspected or discarded and in more defective units reaching the customer. When estimating the savings due to increasing the lot size, these costs should not be ignored. Their cost can be many times greater than the resulting reduction in acceptance sampling costs. The best lot size is the largest lot size still resulting in fairly homogenous lots.

Both random sampling and stratified sampling ensure the sample is representative of the lot as a whole. Sampling only a limited number of cartons can dramatically weaken the protection provided by the sampling plan if the defects appear in groups. Periodic samples allow the selection of samples to be automated. However, care must be taken to assure equal coverage of cavities, nozzles, and so on.

The program SINGLE can also be used to determine sample sizes and criteria for special studies where one wishes to make statements such as "with 99% confidence, the process is producing product that is less than 500 defects per million." Such special studies should be used during process startups when problems are more likely to occur.

Inspection errors can significantly effect the protection provided by a sampling plan. Procedures were given for determining the effect of inspection error on the AQL, LTPD, and OC curve.

Acceptance sampling should provide an incentive for improvement rather than act as a barrier to improvement. It can be used to translate the cost of defectives going to the customer into the more tangible costs of scrap and labor. Acceptance sampling should not be used to try and drive improvement. Instead, reduced inspection and scrap should be the reward for improvement. Neither should acceptance sampling be used as a crutch.

It is not inconsistent to have a goal of zero defects and yet use sampling plans with acceptance numbers other than zero. Zero defects is a goal for producing units. Acceptance sampling dispositions entire lots once they are produced. It has no control over the production of defects or the disposition of individual defects. It can only sentence entire lots. By the time acceptance sampling is performed, a sizable investment in materials and labor has already been made. What one is willing to accept of an already produced lot is logically different from what desires of new lots. Lots should be dispositioned to provide the highest value to the customer.

References

Juran, J. M. (1988). *Juran's Quality Control Handbook*, Fourth Edition, McGraw-Hill, New York, New York.

Squeglia, Nicholas L. (1986). *Zero Acceptance Number Sampling Plans*. Third Edition, American Society for Quality Control, Milwaukee, Wisconsin.

Shingo, Shigeo (1986). *Zero Quality Control: Source Inspection and the Poka-yoke System*. Productivity Press, Cambridge, Massachusetts.

Wheeler, Donald J. and Chambers, David S. (1986). *Understanding Statistical Process Control*. Statistical Process Controls Inc., Knoxville, Tennessee.

6

Double
Sampling Plans

This chapter has three objectives:

- Learn to make accept and reject decisions using double sampling plans.

- Learn to evaluate the protection provided by double sampling plans.

- Learn to select double sampling plans based on the protection they provide.

The single sampling plans covered so far are but one type of sampling plan. Single sampling plans make separate accept/reject decisions for each lot. They can be used to inspect isolated lots as well as series of lots. Single sampling plans make decisions based on a tally of the number of defects or defectives. They can be used with pass/fail data. They can also be used with measurable characteristics by comparing the measurements with specification limits. Single sampling plans are always applicable.

Double sampling plans also make separate accept/reject decisions for each lot and make decisions based on a tally of the number of defects or defectives. They too are always applicable. Double sampling plans differ from single sampling plans in that the number of units inspected varies from lot to lot. Procedurally, double sampling plans are more complex. In return double sampling plans can provide the same protection as single sampling plans while inspecting fewer units. The number of units inspected is typically reduced by 10% to 50%. Double sampling has the added psychological advantage of allowing a second chance.

6.1   Characterizing Double Sampling Plans

Double sampling plans have five parameters:

$n1$ = first sample size
$a1$ = first accept number
$r1$ = first reject number
$n2$ = second sample size
$a2$ = second accept number

Figure 6.1 shows the procedure used by double sampling plans to accept and reject lots of product. Again, one has a choice between tallying either the number of defects or the number of defective units. For higher defect rates, the number of defects should be tallied. When tallying the number of defects, the sample sizes are not limited to integer values.

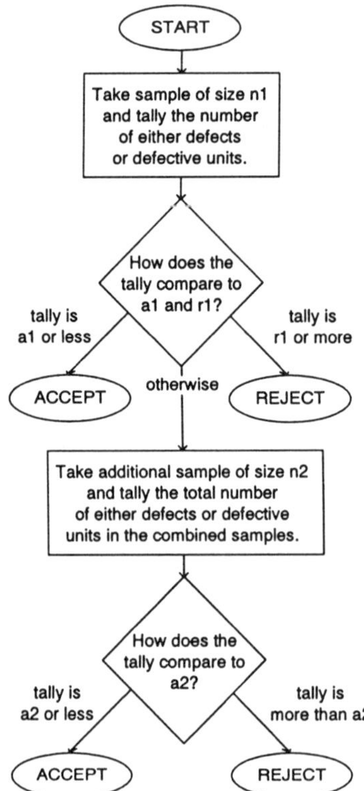

Figure 6.1: Procedure for Double Sampling Plans

Table 6.1 shows the results of applying the double sampling plan n1=32, a1=0, r1=2, n2=32 and a2 =1 to the inspection of truckloads of apples. One should note that n2 is the number of additional units inspected in the second sample. However, a2 should be compared to the cumulative number of defectives in both samples. In the fifth lot, one defective was found in the first sample and one defective in the second sample. The cumulative number of defectives is 1+1=2. Since this is greater than a2, the fifth lot was rejected.

Table 6.1:   Inspection Results

Q. C. Checker Chart					
Product: Apples			Quantity: Truck Load		
Plan: n1=32, a1=0, r1=2, n2=32, a2=1					
Truckload	Number of Defectives 1st Sample	Decision 1st Sample	Number of Defectives 2nd Sample	Decision 2nd Sample	Initials
91-07-1	0	accept			WT
91-07-2	0	accept			WT
91-07-3	2	reject			WT
91-07-4	0	accept			WT
91-07-5	1	2nd sample	1	reject	WT
91-07-6	0	accept			HB
91-07-7	0	accept			HB
91-07-8	1	2nd sample	0	accept	HB
91-07-9	0	accept			HB
91-07-10	0	accept			HB

The double sampling plan used above offers the same protection as the single sampling plan n=50 and a=1. Since two lots required a second sample of 32 units, the average number of units inspected per truckload using the double sampling plan was 38.4. This is a 23% reduction compared to 50 per truckload for the single sampling plan.

Exercises

(6.1)   Make sure you reach the same decisions as in Table 6.1.

(6.2)   Using the same double sampling plan as Table 6.1, what decision would you make if 3 defectives were found in the first sample? How about if 1 defective was found in the first sample and 3 were found in the second sample?

6.2   OC Curves

OC Curves describe the protection provided by acceptance sampling plans. OC curves for double sampling plans can be obtained using the program DOUBLE on the Sampling Programs diskette. The program's main menu is shown in Screen 6.1. Option 3 calculates OC Curves. However, before selecting option 3, a sampling plan must be entered. Option 1 is used to enter this sampling plan. Screen 6.2 shows the use of option 1 to enter the double sampling plan n1=32, a1=0, r1=2, n2=32 and a2=1 where defectives are tallied.

Screen 6.1: Main Menu (DOUBLE)

```
*********************************************************************************
*                              *                                               *
*                              *      Copyright (C) 1992 Taylor Enterprises    *
*      Program DOUBLE          *         P.O. Box 820, Lake Villa, IL 60046    *
*                              *                (708) 356-1074                 *
*                              *                                               *
*********************************************************************************

            Current Sampling Plan:   no plan selected




                      -----  MENU OPTIONS  -----

            (1)  Enter sampling plan
            (2)  Summary information
            (3)  OC curve
            (4)  AOQ curve
            (5)  ASN curve
            (6)  Select sampling plan based on AQL and LTPD


 ENTER NUMBER OF OPTION OR "q" TO QUIT   -->
```

Once the double sampling plan is entered, one can select option 3 to plot or tabulate the OC curve. Screen 6.3 then appears requesting the type of OC curve, whether a plot or table is desired, and information on scaling. If a Type-A OC curve is requested, the lot size is also required. Screen 6.3 shows the input required to obtain a plot of the Type-B OC curve using the default scale. The resulting plot is displayed in Screen 6.4. Screen 6.5 shows the same OC curve in table form.

Screen 6.2: Entering Sampling Plan (DOUBLE)

```
***********************************************************************************
***                                                                            ***
***              DOUBLE Option 1 - Enter Double Sampling Plan                   ***
***                                                                            ***
***********************************************************************************

    Enter the first sample size "n1", first accept number "a1", first reject
    number "r1", second sample size "n2" and second accept number "a2" for the
    desired double sampling plan.  One must also specify whether defectives
    or defects are tallied.  If defects are tallied, one must further specify
    whether the lot consists of individual units or is a continuous lot such as
    a tank of solution or load of cement.  When defects are tallied and the
    lot is continuous, the sample size is not limited to integer values.

 TALLY? (1=defectives,2=defects)            --> 1
 ENTER FIRST SAMPLE SIZE        (n1)        --> 32
 ENTER FIRST ACCEPT NUMBER      (a1)        --> 0
 ENTER FIRST REJECT NUMBER      (r1)        --> 2
 ENTER SECOND SAMPLE SIZE       (n2)        --> 32
 ENTER SECOND ACCEPT NUMBER     (a2)        --> 1
```

Screen 6.3: Entering OC Curve Information (DOUBLE)

```
***********************************************************************************
***                                                                            ***
***                    DOUBLE Option 3 - OC Curve                               ***
***                                                                            ***
***********************************************************************************

   This option plots or tabulates the OC curve.  You must specify which type
   of OC curve to use:

       (A) Type A: describes protection for individual lots
       (B) Type B: describes protection for a series of lots, i.e., a process

   For Type-A OC curves, the lot size must be specified.  One may specify
   the scale or let the program select the appropriate scale.

 ENTER TYPE OF OC CURVE. ("A" or "B")      --> b
 ENTER "p" FOR PLOT "t" FOR TABLE          --> p

   The default scale for percent defective (bottom axis) is from 0%
   to  10.0000000% defective.

 USE DEFAULT SCALE? ("y" or "n")           --> y

   The default scale for probability of acceptance (left axis) is from 0 to 1.

 USE DEFAULT SCALE? ("y" or "n")           --> y
```

Screen 6.4: Plot of Type-B OC Curve (DOUBLE)

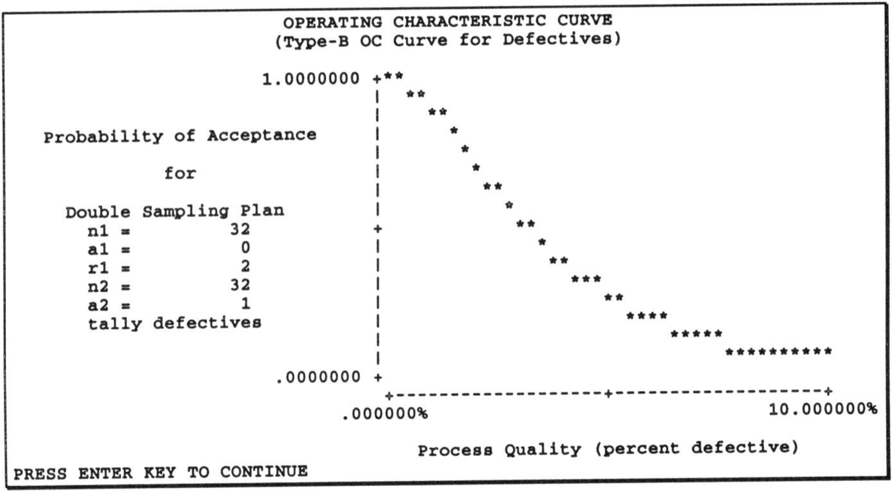

Screen 6.5: Table of Type-B OC Curve (DOUBLE)

```
            OPERATING CHARACTERISTIC CURVE
            (Type-B OC Curve for Defectives)

 Current Sampling Plan:  n1 =       32,   a1 =    0,   r1 =    2
                         n2 =       32,   a2 =    1
                         tally defectives


    Percent Defective            Probability of Acceptance
    _________________            _________________________

         .0000000                        1.00000000
        1.0000000                         .89487040
        2.0000000                         .70311870
        3.0000000                         .51820150
        4.0000000                         .36860960
        5.0000000                         .25690990
        6.0000000                         .17700390
        7.0000000                         .12120810
        8.0000000                         .08276910
        9.0000000                         .05647002
       10.0000000                         .03852887

 PRESS ENTER KEY TO CONTINUE
```

Figure 6.2 shows a plot of the OC curve for this double sampling plan along with that of the single sampling plan n=50 and a=1. These OC curves are close matches. The formulas used to calculate OC curves are given in Appendix B.

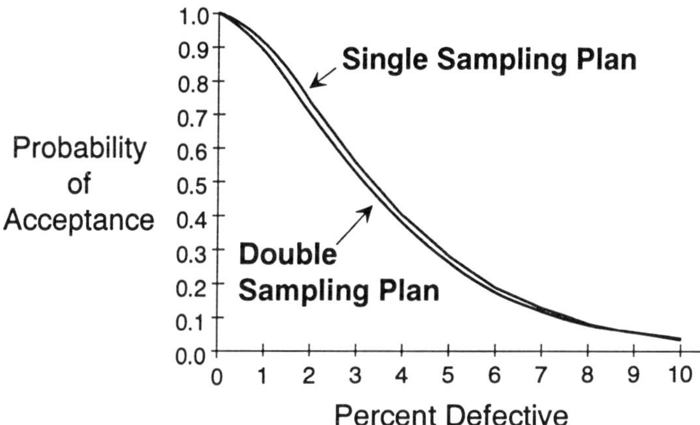

**Figure 6.2: Type-B OC Curves of Single Sampling Plan (n=50, a=1)
and Double Sampling Plan (n1=32, a1=0, r1=2, n2=32, a2=2)**

Exercises

(6.3)   Obtain a table and plot of the Type-A OC curve for the double
sampling plan n1=32, a1=0, r1=2, n2=32 and a2=1 when
tallying defectives and the lot size is 100.   What is the
probability of accepting 1%, 4% and 8% defective lots?

(6.4)   Obtain a table and plot of the Type-A OC curve for the double
sampling plan n1=32, a1=0, r1=2, n2=32 and a2=1 when
tallying defectives and the lot size is 1,000.   What is the
probability of accepting 1%, 4% and 8% defective lots?

(6.5)   Obtain a table and plot of the Type-A OC curve for the double
sampling plan n1=32, a1=0, r1=2, n2=32 and a2=1 when
tallying defectives and the lot size is 10,000.   What is the
probability of accepting 1%, 4% and 8% defective lots?

(6.6)   Compare the OC curves from exercises 6.3 to 6.5 to the one in
Screens 6.4 and 6.5. What is the effect of lot size?

(6.7)   Obtain a table and plot of the Type-B OC curve for the double
sampling plan n1=32, a1=0, r1=2, n2=32, and a2=1 when
tallying defects from a continuous lot. What is the probability
of accepting lots when the process is averaging 0.01, 0.04 and
0.08 defects per unit?

6.3   AQL, IQ, LTPD and AOQL

Option 2 of program DOUBLE can be used to obtain the AQL, IQ, LTPD and AOQL for any double sampling plan. One must specify whether to use the Type-A or Type-B OC curve. The AOQL is only given if a Type-B OC curve is specified. Screen 6.6 shows the summary information for the double sampling plan n1=32, a1=0, r1=2, n2=32 and a2=1 using the Type-B OC curve and assuming defectives are tallied.

This double sampling plan has an AQL of 0.65% and LTPD of 7.50%. In Chapter 2 we saw that the matching single sampling plan, n=50 and a=1, has an AQL of 0.72% and LTPD of 7.56%. While not identical, they are very close. These two sampling plans provide the equivalent protection.

Recall that the AOQL calculated using option 2 assumes the very restrictive assumptions that: rejected lots are 100% inspected, the 100% inspection is 100% effective, the lot size is large compared to the sample size, and that defective units found in either the sample or as part of the 100% inspection are replaced or repaired.

Exercises

(6.8)   What is the AQL, IQ and LTPD of the double sampling plan n1=32, a1=0, r1=2, n2=32 and a2=1 using the Type-A OC curve when defectives are tallied and the lot size is 100?

(6.9)   What is the AQL, IQ and LTPD of the double sampling plan n1=32, a1=0, r1=2, n2=32 and a2=1 using the Type-A OC curve when defectives are tallied and the lot size is 1,000?

(6.10)   What is the AQL, IQ and LTPD of the double sampling plan n1=32, a1=0, r1=2, n2=32 and a2=1 using the Type-A OC curve when defectives are tallied and the lot size is 10,000?

(6.11)   Compare the results of exercises 6.8 to 6.10 to Screen 6.6. What effect does lot size have on the AQL and LTPD?

(6.12)   What is the AQL, IQ, LTPD and AOQL of the double sampling plan n1=32, a1=0, r1=2, n2=32 and a2=1 using the Type-B OC curve when defects are tallied and the lot is continuous?

Screen 6.6: Summary Information (DOUBLE)

```
                        SUMMARY STATISTICS
                  (Type-B OC Curve for Defectives)
   Current Sampling Plan:   n1 =         32,   a1 =     0,   r1 =     2
                            n2 =         32,   a2 =     1
                            tally defectives


                                            Process
                Percentile      Symbol    Percent Defective
                _________       ______    _________________

                  95.0%         AQL              .6469218
                  90.0%                          .9700742
                  50.0%         IQ              3.1090100
                  10.0%         LTPD            7.5047520
                   5.0%                         9.3182300


        AOQL =      1.5548820%    (Assumes 100% inspection of rejected
                                   lots with all defectives being found
                                   and repaired plus a large lot size.)

   PRESS ENTER KEY TO CONTINUE
```

6.4   AOQ Curves

Option 4 of DOUBLE can be used to explore the AOQ under less restrictive conditions. One can obtain either a plot or table of the AOQ curve. One must specify the efficiency of the 100% inspection, the lot size, and the method of handling defectives or defects. The only assumption made is that rejected lots are 100% inspected. Table 6.2 summarizes the different methods of handling defectives or defects found as part of the 100% inspection. Values of the AOQ are calculated using the formulas given in Chapter 2 with the ASN of the double sampling plan substituted for the sample size of the single sampling plan.

Table 6.2: Methods of Handling Defectives and Defects

Case	Methods
Tally Defectives	repair defectives, replace defectives, discard defectives
Tally Defects Lot Contains Units	repair defects, replace defectives, discard defectives
Tally Defects Continuous Lot	remove defects

Screen 6.7 shows the input required for a table of AOQ values when the 100% inspection is 100% efficient, the lot size is large and defectives are replaced. These are the conditions assumed under option 2 in calculating the AOQL. The resulting table is shown in Screen 6.8. A plot of these values is shown in Figure 6.3.

Screen 6.7: Entering AOQ Curve Information (DOUBLE)

```
**************************************************************************
***                                                                  ***
***                  DOUBLE Option 4 - AOQ Curve                     ***
***                                                                  ***
**************************************************************************

   This option plots or tabulates the AOQ curve.  It is assumed that
   rejected lots are 100% inspected.  One must enter the efficiency of
   the 100% inspection, the lot size, and whether defects and
   defectives found are either repaired, replaced or discarded.
   One can specify the scale or let the program select the appropriate scale.

ENTER EFFICIENCY OF 100% INSPECTION      --> 100
ENTER LOT SIZE (0 = Large)               --> 0
DEFECTS? (1=repair,2=replace,3=discard)  --> 2
ENTER "p" FOR PLOT "t" FOR TABLE         --> t

   The default is for the table to start at 0% defective and go up
   to 10.0000000% defective in increments of   1.0000000% defective.

USE DEFAULT SCALE? ("y" or "n")          --> y
```

Screen 6.8: Table of AOQ Curve (DOUBLE)

```
                   AVERAGE OUTGOING QUALITY CURVE
           (N = large, E = 100.000000, replace defectives)

   Current Sampling Plan:  n1 =        32,  a1 =      0,  r1 =      2
                           n2 =        32,  a2 =      1
                           tally defectives


         Percent Defective              Average Outgoing Quality
      ____________________           ____________________________

              .0000000                         .0000000
             1.0000000                         .8948704
             2.0000000                        1.4062370
             3.0000000                        1.5546050
             4.0000000                        1.4744380
             5.0000000                        1.2845500
             6.0000000                        1.0620230
             7.0000000                         .8484569
             8.0000000                         .6621528
             9.0000000                         .5082303
            10.0000000                         .3852880

PRESS ENTER KEY TO CONTINUE
```

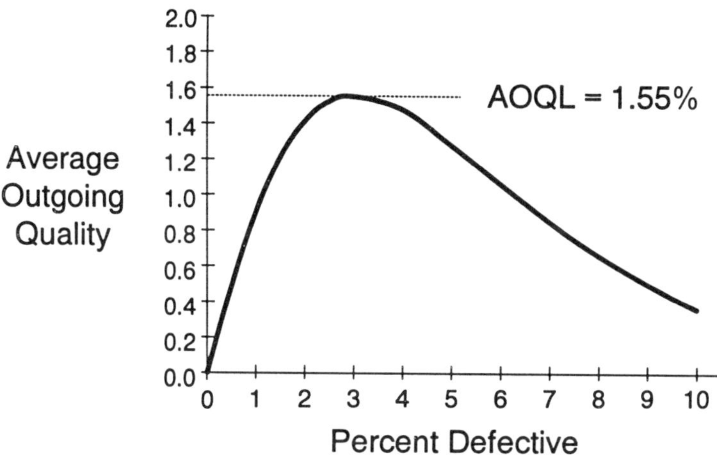

Figure 6.3: AOQ Curve for Double Sampling Plan
n1=32, a1=0, r1=2, n2=32, a2=2

Exercise

(6.13) Obtain a plot and table of the AOQ curve for the double
sampling plan n1=32, a1=0, r1=2, n2=32 and a2=1 when
defectives are replaced, the lot size is large, but the 100%
inspection is only 90% efficient. What is the AOQ at 1%, 4%
and 8% defective? How does this AOQ curve compare to the
one in Figure 6.3?

6.5   ASN Curves

The number of units inspected varies from lot to lot as a result of the
fact that only n1 units are required to accept or reject some lots while
n1+n2 units are required for others. The probability of having to
take the second sample depends on lot quality. Very good lots are
routinely released following the first sample. Very bad lots are
routinely rejected following the first sample. Somewhere in between,
the probability of taking the second sample reaches its maximum.
The ASN curve is a plot of the average number of units inspected
versus lot quality.

As was the case for single sampling plans, accept and reject decisions are sometimes determined before all the units have been inspected. The number of units inspected can therefore be reduced by curtailing the inspection. Inspection can be curtailed on the first and second samples and on acceptance and rejection. The curtailing rules are:

First Sample: Curtail on Rejection: r1 defectives.
                 Curtail on Acceptance: n1-a1 nondefectives.

Second Sample: Curtail on Rejection: a2+1 defectives.
                 Curtail on Acceptance: n1+n2-a2 nondefectives.

One could also choose to curtail only on rejection, only on the second sample, and so on. When defects are tallied, more than one defect per unit can occur. Therefore, inspection can not be curtailed on acceptance. Whether or not curtailing is performed and the type of curtailing effects the ASN curve. In no way does this effect the protection. The AQL, LTPD and OC curve are all unchanged.

Curtailing is not generally performed on the first sample. The first sample is frequently used for other purposes such as control charts or estimating the incoming quality. Curtailing on the first sample interferes with these other uses. The most common form of curtailing is to curtail only on second sample rejection. Curtailing on second sample acceptance offers only minimal benefit while complicating the inspection process.

The program DOUBLE can be used to obtain plots and tables of ASN curves. Assume the double sampling plan n1=32, a1=0, r1=2, n2=32 and a2=1 where defectives are tallied has already been entered. The ASN curve can then be obtained by selecting option 5 from the main menu. The program responds by requesting the method of curtailing and whether a plot or table is desired. Screen 6.9 shows the input required for a plot of the ASN curve under full curtailing. The resulting plot is displayed in Screen 6.10.

Double sampling plans are frequently used in place of single sampling plans to reduce the amount of inspection. Figure 6.2 shows how closely the OC curves of the single sampling plan n=50 and a=1 and the double sampling plan n1=32, a1=0, r1=2, n2=32 and a2=1 match. Figure 6.4 shows ASN curves for these two plans under various methods of curtailing. The double sampling plan requires the inspection of fewer units. Near the AQL, the number of units inspected is reduced approximately 25%.

Screen 6.9: Entering ASN Curve Information (DOUBLE)

```
******************************************************************************
***                                                                        ***
***                   DOUBLE Option 5 - ASN Curve                          ***
***                                                                        ***
******************************************************************************

    This option plots or tabulates the ASN curve.  The method of curtailing must
    be specified.  Curtailing on first sample rejection stops when r1 defects or
    defectives are found.  Curtailing on first sample acceptance stops when
    n1-a1 nondefectives are found.  Curtailing on second sample rejection stops
    when r2+1 defects or defectives are found.  Curtailing on second sample
    acceptance stops when n1+n2-a2 nondefectives are found.  Curtailing on
    acceptance is possible when defects are tallied.  One may specify the
    scale or let the program select the appropriate scale.

CURTAIL 1ST SAMPLE ACCEPT? ("y" or "n") --> y
CURTAIL 1ST SAMPLE REJECT? ("y" or "n") --> y
CURTAIL 2ND SAMPLE ACCEPT? ("y" or "n") --> y
CURTAIL 2ND SAMPLE REJECT? ("y" or "n") --> y
ENTER "p" FOR PLOT "t" FOR TABLE        --> p

    The default scale for percent defective (bottom axis) is from 0%
    to  10.0000000% defective.

USE DEFAULT SCALE? ("y" or "n")             --> y

    The default scale for average sample number (left axis) is from 0
    to        64.00000.

USE DEFAULT SCALE? ("y" or "n")             --> y
```

Screen 6.10: Plot of ASN Curve Under Full Curtailing (DOUBLE)

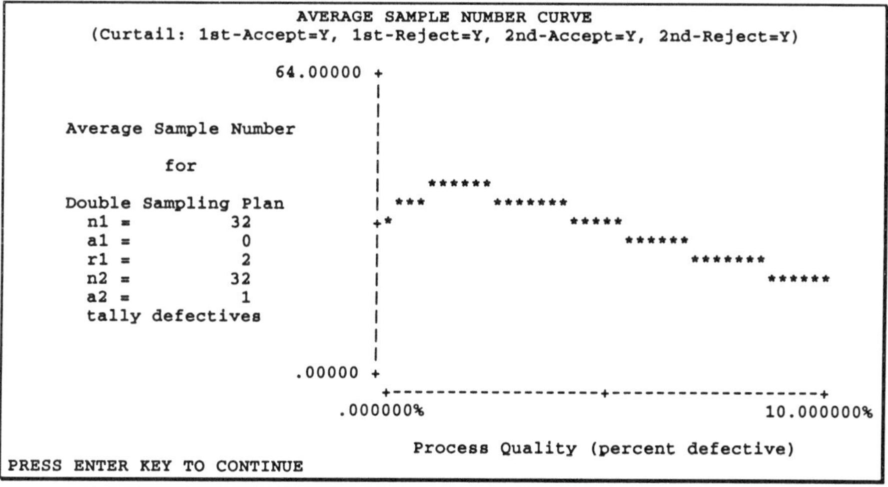

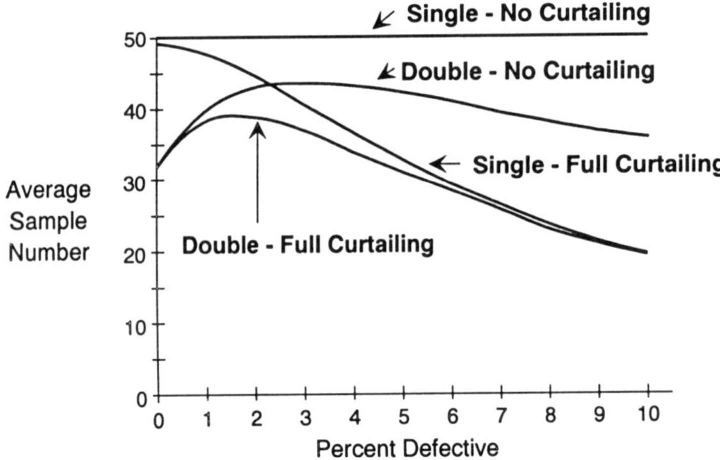

Figure 6.4: ASN Curves for Single Sampling Plan (n=50, a=1) and Double Sampling Plan (n1=32, a1=0, r1=2, n2=32, a2=2)

Double sampling plans also have their disadvantages. The inspection procedure is more complicated. Further, the maximum number of units inspected from a lot is greater: 64 versus 50. This may require setting more units aside. Finally, two samples are required rather than one. Suppose an inspection consists of loading the complete sample on a tray and running it through a sterilizer. A second sample would require a second sterilizer run. It is easier to test all the units in one sterilizer run even if a few more units are required. No one type of sampling plan is best suited for all circumstances. That is why there are so many alternatives.

Up to now we have seen how to evaluate double sampling plans. Options 2, 3, 4 and 5 can be used to obtain the OC, AOQ and ASN curves as well as the AQL, IQ, LTPD and AOQL of any double sampling plan. Next we turn our attention to selecting double sampling plans.

Exercise

(6.14)   Obtain a plot and table of the ASN curve of the double sampling plan with n1=32, a1=0, r1=2, n2=32 and a2=1 when defective units are tallied and inspection is curtailed only on second sample rejection. What is the ASN at 1%, 4% and 8% defective? How does this curve compare to the ASN curves in Figure 6.4?

6.6   Selecting Plans Based on AQL and LTPD

Acceptance sampling plans can provide protection against the release of highly defective lots. To select a sampling plan for this purpose, one must specify an LTPD representing a level of defects to guard against. One also wants to avoid the rejection of good lots. This is accomplished by specifying an AQL. Typically the AQL is above the current process average. Of those double sampling plans having the specified AQL and LTPD, the one that minimizes the amount of inspection is desired.

The program DOUBLE can be used to select double sampling plans on the basis of their AQL and LTPD. To select a plan, select option 6 from the main menu. The program will request the desired AQL and LTPD along with whether defectives or defects are tallied and the type of OC curve to use. One must also specify a process or lot quality p_0 at which to minimize the ASN. The double sampling plan will be selected that minimizes the ASN at p_0 from those satisfying the conditions:

$$\text{Actual AQL} \;\geq\; \text{Specified AQL}$$

$$\text{Actual LTPD} \;\leq\; \text{Specified LTPD}$$

One must also specify whether to use the standard definitions of the AQL (95%) and LTPD (10%).

Suppose one wishes to select the double sampling plan with an AQL of 1% and an LTPD of 10% that minimizes the ASN at 0.5%. This sampling plan is to be based on the Type-B OC for defectives. Screen 6.11 shows the required input. After the last question is answered, the program displays a message that it is searching for a plan. Depending on the speed of your computer and the values entered, it can take from a fraction of a second up to several hours to determine the plan. Once the desired double sampling plan has been found, the program returns to the main menu with the selected plan displayed at the top (Screen 6.12). The selected plan is n1=23, a1=0, r1=2, n2=46 and a2=2. The algorithms used by DOUBLE are given in Taylor (1983) and Taylor (1986).

An alternate approach to selecting double sampling plans with specified AQLs and LTPDs is to use Table 6.3. Table 6.3 provides double sampling plans indexed by their AQL and LTPD. This table assumes defectives are tallied and is based on the Type-B OC curve. It covers the same range of AQLs and LTPDs as Table 3.3 for single

sampling plans. The double sampling plans in Table 6.3 minimize
the ASN at the AQL. The parameters of the double sampling plans
are represented as follows:

$$n1/(a1,r1)$$
$$n2/(a2,a2+1)$$

Screen 6.11: Selecting Plan Based on AQL and LTPD (DOUBLE)

```
**********************************************************************************
***                                                                          ***
***        DOUBLE Option 6 - Select Sampling Plan Based on AQL and LTPD       ***
***                                                                          ***
**********************************************************************************

    This option selects the double sampling plan minimizing the ASN at a
    specified point P0 from among those satisfying the following conditions:

        Actual AQL is greater than or equal to the specified AQL
        Actual LTPD is less than or equal to the specified LTPD

    One must specify whether to use the Type-A or Type-B OC curve for
    defects or defectives.  The standard definitions of AQL (95% chance of
    acceptance) and LTPD (10% chance of acceptance) are used unless indicated
    otherwise.

TALLY? (1=defectives,2=defects)          --> 1
ENTER TYPE OF OC CURVE. ("A" or "B")     --> b
ENTER AQL  AS PERCENT DEFECTIVE          --> 1
ENTER LTPD AS PERCENT DEFECTIVE          --> 10
ENTER P0   AS PERCENT DEFECTIVE          --> 0.5
USE .95 & .1 FOR AQL-LTPD? ("y" or "n") --> y

        Please Wait: Selecting double sampling plan.
```

Screen 6.12: Selected Sampling Plan (DOUBLE)

```
**********************************************************************************
*                              *                                               *
*                              *        Copyright (C) 1992 Taylor Enterprises  *
*      Program DOUBLE          *          P.O. Box 820, Lake Villa, IL 60046   *
*                              *                 (708) 356-1074                *
*                              *                                               *
**********************************************************************************

      Current Sampling Plan:  n1 =         23,   a1 =      0,   r1 =      2
                              n2 =         46,   a2 =      2
                              tally defectives


                          ----- MENU OPTIONS -----

              (1) Enter sampling plan
              (2) Summary information
              (3) OC curve
              (4) AOQ curve
              (5) ASN curve
              (6) Select sampling plan based on AQL and LTPD


ENTER NUMBER OF OPTION OR "q" TO QUIT  -->
```

Table 6.3: Double Sampling Plans for Defectives by AQL and LTPD

AQL	Approximate Ratio of LTPD / AQL								
	45	11	6.5	5	4	3.2	2.8	2.3	2
10%	-	3/(0,2) 1/(1,2) AQL = 9.76 LTPD = 68.0 AOQL = 19.8	5/(0,3) 4/(2,3) AQL = 9.94 LTPD = 50.0 AOQL = 15.4	5/(0,3) 10/(3,4) AQL = 9.95 LTPD = 42.9 AOQL = 13.6	10/(1,4) 16/(5,6) AQL = 10.7 LTPD = 36.3 AOQL = 12.4	14/(1,5) 20/(6,7) AQL = 10.1 LTPD = 30.2 AOQL = 11.6	25/(3,8) 42/(10,11) AQL = 10.0 LTPD = 25.5 AOQL = 10.6	40/(4,10) 55/(14,15) AQL = 10.1 LTPD = 21.3 AOQL = 10.2	50/(4,14) 90/(20,21) AQL = 10.3 LTPD = 19.1 AOQL = 10.1
6.5%	-	4/(0,2) 2/(1,2) AQL = 6.50 LTPD = 52.7 AOQL = 13.9	6/(0,3) 8/(2,3) AQL = 6.49 LTPD = 37.1 AOQL = 10.4	8/(0,3) 14/(3,4) AQL = 6.54 LTPD = 30.2 AOQL = 9.19	16/(1,5) 28/(5,6) AQL = 6.53 LTPD = 23.5 AOQL = 7.87	26/(2,6) 40/(7,8) AQL = 6.55 LTPD = 20.0 AOQL = 7.36	45/(3,8) 65/(11,12) AQL = 6.53 LTPD = 15.7 AOQL = 6.81	70/(5,10) 100/(16,17) AQL = 6.51 LTPD = 13.8 AOQL = 6.56	110/(8,15) 140/(22,23) AQL = 6.49 LTPD = 12.2 AOQL = 6.37
4.0%	-	6/(0,2) 4/(1,2) AQL = 3.95 LTPD = 36.7 AOQL = 8.82	10/(0,3) 12/(2,3) AQL = 4.04 LTPD = 24.7 AOQL = 6.62	14/(0,3) 20/(3,4) AQL = 4.05 LTPD = 19.7 AOQL = 5.81	26/(1,4) 40/(5,6) AQL = 4.06 LTPD = 15.4 AOQL = 5.04	45/(2,5) 50/(7,8) AQL = 3.98 LTPD = 12.7 AOQL = 4.65	70/(3,8) 90/(10,11) AQL = 4.00 LTPD = 10.2 AOQL = 4.24	100/(4,11) 155/(15,16) AQL = 4.05 LTPD = 8.70 AOQL = 4.09	145/(6,13) 230/(21,22) AQL = 4.05 LTPD = 7.81 AOQL = 3.99
2.5%	*	8/(0,2) 8/(1,2) AQL = 2.60 LTPD = 27.0 AOQL = 6.02	16/(0,3) 20/(2,3) AQL = 2.46 LTPD = 15.9 AOQL = 4.10	20/(0,3) 36/(3,4) AQL = 2.52 LTPD = 12.9 AOQL = 3.65	45/(1,4) 55/(5,6) AQL = 2.57 LTPD = 9.67 AOQL = 3.18	55/(1,6) 80/(6,7) AQL = 2.53 LTPD = 8.21 AOQL = 2.93	115/(3,9) 140/(10,11) AQL = 2.50 LTPD = 6.37 AOQL = 2.65	155/(4,12) 225/(14,15) AQL = 2.51 LTPD = 5.59 AOQL = 2.55	250/(7,13) 350/(21,22) AQL = 2.52 LTPD = 4.96 AOQL = 2.49
1.5%	*	13/(0,2) 13/(1,2) AQL = 1.60 LTPD = 17.5 AOQL = 3.77	25/(0,3) 35/(2,3) AQL = 1.49 LTPD = 10.1 AOQL = 2.51	30/(0,3) 65/(3,4) AQL = 1.52 LTPD = 8.25 AOQL = 2.22	70/(1,5) 65/(4,5) AQL = 1.53 LTPD = 6.31 AOQL = 1.98	80/(1,5) 150/(6,7) AQL = 1.50 LTPD = 5.20 AOQL = 1.76	145/(2,9) 230/(9,10) AQL = 1.50 LTPD = 4.08 AOQL = 1.62	240/(4,9) 335/(13,14) AQL = 1.50 LTPD = 3.54 AOQL = 1.56	360/(6,12) 540/(19,20) AQL = 1.50 LTPD = 3.08 AOQL = 1.50
1.0%	*	20/(0,2) 20/(1,2) AQL = 1.04 LTPD = 11.8 AOQL = 2.47	45/(0,3) 40/(2,3) AQL = 1.01 LTPD = 6.56 AOQL = 1.68	55/(0,3) 80/(3,4) AQL = 1.00 LTPD = 5.25 AOQL = 1.47	110/(1,4) 145/(5,6) AQL = 1.00 LTPD = 3.94 AOQL = 1.26	140/(1,6) 200/(6,7) AQL = 0.997 LTPD = 3.30 AOQL = 1.16	290/(3,9) 340/(10,11) AQL = 1.01 LTPD = 2.58 AOQL = 1.07	450/(5,10) 580/(15,16) AQL = 0.999 LTPD = 2.21 AOQL = 1.01	625/(7,14) 900/(21,22) AQL = 1.00 LTPD = 1.98 AOQL = 0.987
0.65%	*	32/(0,2) 32/(1,2) AQL = 0.647 LTPD = 7.50 AOQL = 1.55	82/(0,3) 46/(2,3) AQL = 0.652 LTPD = 4.21 AOQL = 1.09	85/(0,4) 135/(3,4) AQL = 0.652 LTPD = 3.33 AOQL = 0.933	160/(1,5) 270/(5,6) AQL = 0.652 LTPD = 2.57 AOQL = 0.803	270/(2,6) 370/(7,8) AQL = 0.652 LTPD = 2.09 AOQL = 0.742	460/(3,10) 630/(11,12) AQL = 0.653 LTPD = 1.62 AOQL = 0.684	700/(5,11) 1000/(16,17) AQL = 0.652 LTPD = 1.42 AOQL = 0.658	1000/(7,14) 1600/(23,24) AQL = 0.648 LTPD = 1.25 AOQL = 0.637
0.4%	*	50/(0,2) 50/(1,2) AQL = 0.414 LTPD = 4.87 AOQL = 0.999	120/(0,3) 90/(2,3) AQL = 0.402 LTPD = 2.64 AOQL = 0.675	130/(0,4) 230/(3,4) AQL = 0.401 LTPD = 2.10 AOQL = 0.577	300/(1,4) 300/(5,6) AQL = 0.397 LTPD = 1.58 AOQL = 0.511	425/(2,6) 625/(7,8) AQL = 0.401 LTPD = 1.31 AOQL = 0.458	700/(3,9) 900/(10,11) AQL = 0.399 LTPD = 1.04 AOQL = 0.424	1100/(5,11) 1500/(15,16) AQL = 0.398 LTPD = 0.892 AOQL = 0.404	-

* Use single sampling plan from Table 3.3 (page 47).

Table 6.3: (continued)

AQL	Approximate Ratio of $LTPD/AQL$								
	45	11	6.5	5	4	3.2	2.8	2.3	2
0.25%	*	80/(0,2) 80/(1,2) AQL = 0.258 LTPD = 3.07 AOQL = 0.626	170/(0,3) 170/(2,3) AQL = 0.253 LTPD = 1.70 AOQL = 0.427	200/(0,4) 380/(3,4) AQL = 0.251 LTPD = 1.34 AOQL = 0.362	450/(1,5) 550/(5,6) AQL = 0.250 LTPD = 0.994 AOQL = 0.318	550/(1,6) 800/(6,7) AQL = 0.251 LTPD = 0.839 AOQL = 0.293	1200/(3,9) 1300/(10,11) AQL = 0.252 LTPD = 0.644 AOQL = 0.267	.	.
0.15%	*	125/(0,2) 125/(1,2) AQL = 0.165 LTPD = 1.97 AOQL = 0.401	240/(0,3) 360/(2,3) AQL = 0.149 LTPD = 1.07 AOQL = 0.256	315/(0,3) 630/(3,4) AQL = 0.150 LTPD = 0.832 AOQL = 0.221	700/(1,5) 675/(4,5) AQL = 0.150 LTPD = 0.637 AOQL = 0.196	800/(1,5) 1500/(6,7) AQL = 0.149 LTPD = 0.529 AOQL = 0.176	.	.	.
0.1%	*	200/(0,2) 200/(1,2) AQL = 0.103 LTPD = 1.24 AOQL = 0.251	480/(0,3) 360/(2,3) AQL = 0.100 LTPD = 0.665 AOQL = 0.169	500/(0,4) 950/(3,4) AQL = 0.100 LTPD = 0.536 AOQL = 0.145	1100/(1,4) 1400/(5,6) AQL = 0.101 LTPD = 0.403 AOQL = 0.128	.	.	.	.
0.065%	*	315/(0,2) 315/(1,2) AQL = 0.0656 LTPD = 0.788 AOQL = 0.160	700/(0,3) 650/(2,3) AQL = 0.0633 LTPD = 0.426 AOQL = 0.107	900/(0,4) 1300/(3,4) AQL = 0.0645 LTPD = 0.331 AOQL = 0.0926	.	.	.	.	.
0.04%	*	500/(0,2) 500/(1,2) AQL = 0.0413 LTPD = 0.497 AOQL = 0.101	1100/(0,3) 1100/(2,3) AQL = 0.0390 LTPD = 0.265 AOQL = 0.0661	.	.	.	.	.	.
0.025%	*	800/(0,2) 800/(1,2) AQL = 0.0260 LTPD = 0.311 AOQL = 0.0629	.	.	.	.	.	.	.
0.015%	*	1250/(0,2) 1250/(1,2) AQL = 0.0165 LTPD = 0.199 AOQL = 0.0402	.	.	.	.	.	.	.
0.01% to 0.0025%	*	.	.	.	.	.	.	.	.

* Use single sampling plan from Table 3.3 (page 47).

In some cases Table 6.3 indicates to use the corresponding single sampling plan. This occurs whenever the corresponding single sampling plan has an accept number of zero. In this case, there is no double sampling plan with the same protection that requires fewer samples.

When the program DOUBLE is used to select a double sampling plan with an AQL of 1.0% and LTPD of 10%, the resulting plan is n1=23, a1=0, r1=2, n2=46 and a2=2. The closest plan to the stated conditions in Table 6.3 is n1=20, a1=0, r1=2, n2=20 and a2=1. This plan is found in the row labeled AQL=1.0% and column labeled LTPD/AQL=11. The LTPD of the plan from the table is slightly larger than 10%. However, it is close.

Table 6.3 is based on the Type-B OC curve. These plans can, however, also be used for Type-A OC curves. As was the case for single sampling plans, Type-A OC curves are steeper than the corresponding Type-B OC curves. Double sampling plans selected based on Type-B OC curves satisfying

$$\text{Actual AQL} \geq \text{Specified AQL}$$

$$\text{Actual LTPD} \leq \text{Specified LTPD}$$

will also satisfy these same conditions for all Type-A OC curves regardless of the lot size. Therefore, any plan selected on the basis of Table 6.3, can be used for both processes (Type B) and individual lots (Type A) of any lot size.

When inspecting an isolated lot so that a Type-A OC curve is appropriate, it may be advantageous to select a double sampling plan specific to the Type-A OC curve. This can be accomplished using the program DOUBLE. DOUBLE will ask for the lot size. Further, the values entered for the AQL, LTPD, and p_0 are restricted to the values $100/N$, $200/N$, ..., $(N-1)100/N$. These correspond to 1 defective in the lot, 2 defectives and so on. Table 6.4 shows the result of selecting sampling plans with an AQL of 1.0% and LTPD of 10% that minimize the ASN at 0.5%. The advantage of selecting plans specific to the Type-A OC curve is that the ASN may be reduced. The reduction is generally minimal.

Table 6.3 can also be used to select double sampling plans when defects are tallied so long as the defect rate is low ($< .1$). Otherwise, the program DOUBLE should be used. When using DOUBLE for defects, the AQL, LTPD and p_0 must be given in terms of the average number of defects per unit. Further, DOUBLE will ask if the lot is continuous or consists of units of product. For continuous lots, non integer samples sizes are possible. One must specify the smallest sample size increment n_{inc}. DOUBLE will then restrict the sample sizes to the values: n_{inc}, $2n_{inc}$, $3n_{inc}$, $4n_{inc}$, ⋯. Selecting $n_{inc}=1$ insures integer sample sizes. Screen 6.13 shows the input required to select the double sampling plan for defects corresponding to those in Table 6.4. It is assumed the lots are continuous and that the plan is restricted to integer sample sizes. The resulting plan is shown in Table 6.4 in the row labeled "Type B, Defects". Since the defect rate is low, the selected plan does not differ greatly from the ones for defectives.

Table 6.4: Effect of Lot Size for Type-A OC Curves for Defectives When AQL=1%, LTPD=10% and p_0=0.5%

Lot Size	Sampling Plan				
	n1	a1	r1	n2	a2
100	21	0	2	20	1
200	22	0	2	24	1
500	25	0	2	17	1
1,000	23	0	2	42	2
10,000	23	0	2	46	2
100,000	23	0	2	46	2
Type-B, Defectives	23	0	2	46	2
Type-B, Defects	24	0	2	50	2

Screen 6.13: Selecting Sampling Plan for Defects (DOUBLE)

```
************************************************************************
***                                                                ***
***       DOUBLE Option 6 - Select Sampling Plan Based on AQL and LTPD  ***
***                                                                ***
************************************************************************

    This option selects the double sampling plan minimizing the ASN at a
    specified point P0 from among those satisfying the following conditions:

        Actual AQL is greater than or equal to the specified AQL
        Actual LTPD is less than or equal to the specified LTPD

    One must specify whether to use the Type-A or Type-B OC curve for
    defects or defectives.  The standard definitions of AQL (95% chance of
    acceptance) and LTPD (10% chance of acceptance) are used unless indicated
    otherwise.

TALLY? (1=defectives,2=defects)            --> 2
LOT? (1=units, 2=continuous)               --> 2

    Non integer sample sizes are possible.  You must specify the smallest
    possible increment for the sample size, say x.  Then the possible samples
    sizes are x, 2x, 3x, ... .

ENTER SMALLEST SAMPLE SIZE INCREMENT    --> 1

        Info: current program can only select double sampling plans
              for Type-B OC curves when defects are tallied.

ENTER AQL  AS DEFECT RATE                  --> .01
ENTER LTPD AS DEFECT RATE                  --> 0.1
ENTER P0   AS DEFECT RATE                  --> 0.005
USE .95 & .1 FOR AQL-LTPD? ("y" or "n") --> y
```

Exercises

(6.15) Select the double sampling plan with an AQL of 0.4% and
        LTPD of 5% which minimizes the ASN at 0.4%. Use the
        Type-B OC curve.

(6.16) Select the double sampling plan with an AQL of 0.4% and
        LTPD of 5% which minimizes the ASN at 0.01%. Use Type-B
        OC curve.

(6.17) Select the double sampling plan with an AQL of 10.0 defects
        per unit and LTPD of 50.0 defects per unit which minimizes
        the ASN at 10.0 defects per unit. Use the Type-B OC curve.
        Assume lots are continuous and specify a smallest sample
        size increment of 0.01.

6.7   Selecting Matching Double Sampling Plans

Acceptance sampling plans can also be used to assure the average outgoing quality is below some specified value or to maximize customer value. The program DOUBLE will not directly select sampling plans for these objectives. However, one can always select the appropriate single sampling plan, determine its AQL and LTPD and then select a matching double sampling plan.

If two sampling plans have the same AQL and LTPD, they will have nearly identical OC curves. They will accept the same percentage of 1% defective lots. They will reject the same percentage of 10% defective lots. Their AOQ curves and AOQLs will be the same. Both sampling plans should provide nearly the same customer value. Everything will be the same except for the cost of the sampling inspection. Therefore, selecting a single sampling plan meeting certain requirements and then selecting a matching double sampling plan should result in a double sampling plan that also meets these requirements.

As an example, suppose one wants to select the double sampling plan with an AQL of 2.0 and AOQL of 3.0 defects per unit minimizing the ASN at 1.0 defects per unit. Assume the lot is continuous and that the smallest sample size increment is 0.1. Using the program SINGLE, the best single sampling plan matching these requirements is n=0.9 and a=4. Its actual AQL is 2.19 and actual AOQL is 2.83 defects per unit. Its LTPD is 8.88 defects per unit. Next, the program DOUBLE can be used to select a matching double sampling plan. Specifying an AQL of 2.0, LTPD of 8.9, and p_0=1.0, the matching double sampling plan is n1=0.3, a1=0, r1=3, n2=0.9 and a2=5. This plan has an actual AQL of 2.05 and AOQL of 2.68. This meets the original objective. Trying larger values of the LTPD results in n1=0.2, a1=0, r1=3, n2=0.8, a2=4 with an AQL of 21.3 and AOQL of 2.90 (LTPD=12.0). In the case where defectives are tallied, Table 6.3 can also be used to select double sampling plans with a specified AQL and AOQL.

Exercise:

(6.18)   Select the double sampling plan with an AQL of 1% and AOQL of 2% defective minimizing the ASN at 1% defective. What is the actual AQL and AOQL of the selected plan?

6.8   Reinspections

In some companies it is accepted practice to reinspect rejected lots. Generally a tighter sampling plan is used for reinspection. If Mil-Std-105E is being used (covered in Chapter 8), the original sampling inspection might be a S-3 inspection and the reinspection might be a L-II.

As an example, suppose the single sampling plan n=13 and a=0 is used to initially inspect all lots and that rejected lots are reinspected using the sampling plan n=315 and a=7. It is also common practice to include the first sample as part of the reinspection samples. Such reinspections are essentially home grown double sampling plans. In the example, the procedure used is essentially the double sampling plan n1=13, a1=0, $2 \leq r1 \leq 8$, n2=302 and a2=7. One should question the inspector to find out r1. Assume that r1=2 so that only lots with a single defective in their initial sample qualify for reinspection.

The double sampling plan n1=13, a1=0, r1=2, n2=302 and a2=7 has an AQL of 1.56% and LTPD of 16.2%. At 1% defective it has an ASN of 47.8, assuming no curtailing. Such home grown double sampling plans are frequently very inefficient. Using the program DOUBLE and specifying that AQL=1.5%, LTPD=16.2% and p_0=1%, results in the double sampling plan n1=15, a1=0, r1=2, n2=11 and a2=1. This sampling plan has an actual AQL of 1.52% and LTPD of 16.2%. It provides nearly identical protection. However, the ASN at 1% defective is only 16.4. This is a 66% reduction.

Anytime home grown double sampling plans or reinspection procedures are used, they should be carefully evaluated. Generally, they can be replaced by more efficient double sampling plans. A plant that uses such reinspections can reduce their inspection costs by 10 to 40 percent without changing the protection provided. One should also examine the justification for the reinspection along with the protection provided by the reinspection. Such reinspections are symptomatic of problems in the manner the original plans where selected. Chapter 8 gives an example of how the use of Mil-Std-105E has resulted in many facilities establishing reinspections. Such reinspections may not fully have the desired effect. In replacing these reinspections, the protection provided by the reinspection should not simply be matched.

6.9   Multiple and Sequential Sampling Plans

Double sampling plans require two sets of samples. Sampling plans can also be devised requiring three or more sets of samples. Such sampling plans are called multiple sampling plans. At the extreme are sampling plans that make accept/reject decisions following each unit inspected and allow for the possibility of hundreds of units being inspected. Such sampling plans are called sequential sampling plans. Both multiple and sequential sampling plans can be used to reduce the ASN further; however, at the expense of increased administrative complexity.

Multiple and sequential sampling plans are not covered in this book. Partially, because the amount of reduction in ASN they offer is small. But the second reason is that the procedures for selecting multiple and sequential sampling plans are not as well developed as the procedure used by this book to select double sampling plans. As a result many of the multiple and sequential sampling plans in use are not as efficient as these double sampling plans[1].

Table 6.5 shows matching single, double and multiple sampling plans from Mil-Std-105E. These three plans have AQLs near 1.3% and LTPDs near 4.6% defective. Also given is an alternative double sampling plan selected using the program DOUBLE. This alternative double sampling plan clearly out performs the Mil-Std-105E double sampling plan. In fact, it is comparable in efficiency to the multiple sampling plan.

The reason that the double sampling plans selected by DOUBLE performs so well is that they are selected without any restrictions on the five parameters. Most double and multiple sampling plans given elsewhere place certain restrictions on the parameters. Mil-Std-105E restricts n1=n2 for double sampling plans. As a result, it is frequently possible to use DOUBLE to select double sampling plans more efficient than those given in most tables.

---

[1]For those familiar with sequential sampling plans, Wald seqential sampling plans minimize the ASN at the AQL and LTPD. No double sampling plan can offer better ASNs at these two points. However, what if the process average is half of the AQL. It may be possible to select a double sampling plan more efficient than the Wald's sequential plan at this point.

Table 6.5: Alternative Sampling Plans with
AQL=1.3% and LTPD=4.6%

Source	Sampling Plan	AQL	LTPD	ASN(0)	ASN(.5)	ASN(1)
Mil-Std-105E	Single n=20, a=5	1.3%	4.6%	200.0	200.0	200.0
Mil-Std-105E	Double n1=50, a1=2, r1=5 n2=125, a2=6	1.3%	4.6%	125.0	128.1	140.2
Mil-Std-105E	Multiple n1=50, a1=#, r1=4 n2=50, a2=1, r2=5 n3=50, a3=2, r3=6 n4=50, a4=3, r4=7 n5=50, a5=5, r5=8 n6=50, a6=7, r6=9 n7=50, a7=9	1.4%	4.5%	100.0	106.6	125.8
DOUBLE minimize ASN(1%)	Double n1=91, a1=1, r1=5 n2=170, a2=0	1.3%	4.6%	91.0	104.0	129.9

6.10   Summary

Double sampling plans, like single sampling plans, are always applicable. They make separate accept/reject decisions for each lot. They can be used to inspect isolated lots as well as series of lots. Double sampling plans can offer the same protection as single sampling plans while inspecting fewer units. The number of units inspected is typically reduced by 10% to 50%. However, this savings comes at the expense of increased administrative complexity.

OC curves, AQLs, LTPDs, AOQ curves, AOQLs and ASN curves can all be generated for double sampling plans. Type-A and Type-B OC curves are possible. Inspection can be curtailed on both the first and second samples.

Procedures are given for selecting double sampling plans based on their AQL and LTPD. For single sampling, there exists one single sampling plan satisfying these conditions that minimizes the ASN curve at all points. For double sampling, different double sampling plans minimize the ASN curve at different points. Therefore, one must specify the point on the ASN curve to minimize. This procedure can be used to generate matching double sampling plans.

Whenever reinspections are performed, one is essentially using a double sampling plan. Enormous cost savings are possible by replacing reinspection procedures with more efficient double sampling plans. As part of replacing these reinspections, the justification for reinspection should be reviewed.

The procedure used in this book to select double sampling plans generates double sampling plans that are extremely efficient. No restrictions are placed on the five parameters. These double sampling plans are more efficient than those found in most tables including Mil-Std-105E. In fact, they are frequently as or more efficient than multiple and sequential sampling plans given in such tables.

References:

Taylor, Wayne A. (1986). "A Program for Selecting Efficient Binomial Double Sampling Plans." Journal of Quality Technology, Vol. 18, No. 1, pp 67-73.

Taylor, Wayne A. (1983). *Selecting Efficient Single Stage and Double Stage Attribute Sampling Plans of a Given Power.* Ph.D. Thesis, Purdue University.

7

Quick
Switching Systems

This chapter has three objectives:

- Learn to make accept and reject decisions using QSSs.

- Learn to evaluate the protection provided by QSSs.

- Learn to select QSSs based on the protection they provide.

A quick switching system (QSS) consists of two sampling plans along with a set of rules for switching between them. The first sampling plan, called reduced inspection, is intended for use when the process quality is high. It has a lower sample size to reduce inspection costs. The second sampling plan, called tightened inspection, is intended for use when problems are encountered. It is designed to give a high level of protection. The switching rules are designed to ensure the correct plan is used. They are also designed to be easy to use and to react quickly. QSSs are only applicable when inspecting a series of lots.

The idea of switching between different sampling plans has enormous intuitive appeal. First, it is only common sense to concentrate inspection effort where it is needed most. Quick switching systems do just this by lowering inspection costs until a problem is detected and then increasing the inspection performed until the problem is fixed. Further, a frequent problem when acceptance sampling is the sandwich effect: two lots are rejected but the lot sandwiched between them is accepted. An astute inspector will perform additional inspection on this released lot since it is likely that it too is bad. Switching procedures do this automatically.

7.1   Characterizing Quick Switching Systems

Quick switching systems consist of two sampling plans. The first plan, called reduced inspection, can be either a single or double sampling plan. Its parameters are:

<u>Single Sampling Plan</u>

n_r = sample size

a_r = accept number

<u>Double Sampling Plan</u>

$n1_r$ = first sample size

$a1_r$ = first accept number

$r1_r$ = first reject number

$n2_r$ = second sample size

$a2_r$ = second accept number

Chapters 2 and 6 describe how to make accept and reject decisions for single and double sampling plans.

The second sampling plan, called tightened inspection, is a single sampling plan. Its parameters are:

n_t = sample size

a_t = accept number

In addition, there is one additional parameter called the switch number:

s_t = switch number from tightened to reduced

The switch number is used to determine which sampling plan to use. The first lot inspected using a QSS should be inspected using the tightened inspection. Following each lot inspected, two decisions must be made. The first decision is whether to accept or reject the current lot. The second decision is which of the two sampling plans to use to inspect the next lot. When using the tightened inspection, you switch to reduced if s_t or fewer defects or defectives are found. When using the reduced inspection, you switch to tightened if the lot is rejected. The switching rules are summarized in Figure 7.1. Following a major shutdown, you may decide to restart the QSS using the tightened inspection. There is no need to restart in tightened inspection following breaks, shift changes, and so on unless past experience indicates such events frequently result in problems.

QSSs are procedurally more complex than single or double sampling plans. However, with a well-designed checker chart, they are not a problem to administrate. The offsetting savings in inspection effort is more than worth the effort.

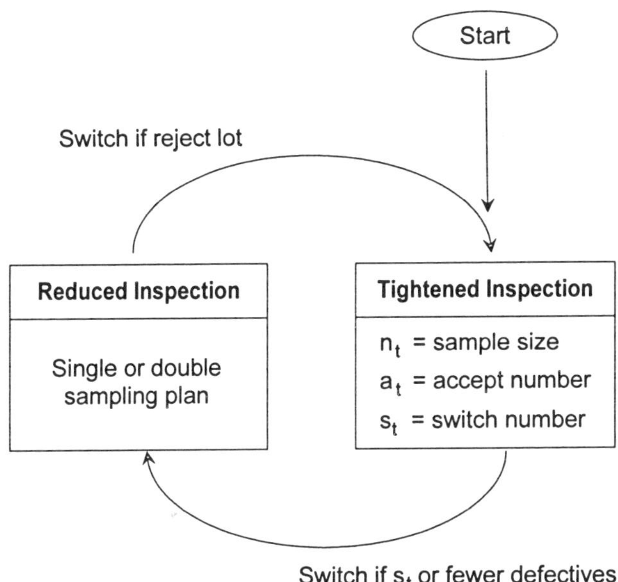

Figure 7.1: Switching Rules for QSS

Table 7.1 shows a checker chart specially designed for the QSS $n1_r{=}18$, $a1_r{=}0$, $r1_r{=}2$, $n2_r{=}80$, $a2_r{=}1$, $n_t{=}50$, $a_t{=}1$ and $s_t{=}0$. This chart shows all the possible outcomes. It is not representative of what one finds in practice. In practice, most of the time is spent in reduced inspection with no defectives being found. This results in the routine selection of 18 samples per lot. The chart then provides instructions for handling less frequent events.

The QSS in Table 7.1 offers the same protection as the single sampling plan n=50 and a=1 and the double sampling plan n1=32, a1=0, r1=2, n2=32 and a2=1. During periods of good quality, this QSS averages close to 18 samples per lot. In this case, the QSS offers a 64% reduction in sample size compared to the single sampling plan and a 44% reduction in sample size compared to the double sampling plan. Such savings more than offset the increased complexity.

QSSs have a further psychological advantage over single and double sampling plans. QSSs are perceived as reacting more severely to problems. They not only reject bad lots, but they tighten up the inspection of future lots. This frequently leads to more immediate corrective action. On some occasions, the replacement of single and double sampling plans with QSSs has resulted in dramatic improvements in the quality coming into an inspection.

Table 7.1: Inspection Results

Q. C. Checker Chart					
Product: Apples			Quantity: Truckload		
Plan: $n1_r=18$, $a1_r=0$, $r1_r=2$, $n2_r=80$, $a2_r=1$, $n_t=50$, $a_t=1$, and $s_t=0$					
Truckload	Reduced Inspection			Tightened Inspection	Decision
	Stage 1	Stage 2	Total Stages 1 and 2		
	Pull 18 samples 0 = A 1 = go stage 2 2 = R,S	Pull 80 samples go to next column	Sum 1 & 2 1 or less = A 2 or more = R,S	Pull 50 samples 0 = A,S 1 = A 2 or more = R	A = Accept R = Reject S = Switch
91-07-1				2	R
91-07-2				1	A
91-07-3				0	A,S
91-07-4	0				A
91-07-5	1	0	1		A
91-07-6	2				R,S
91-07-7				0	A,S
91-07-8	0				A
91-07-9	0				A
91-07-10	1	1	2		R,S

Exercises

(7.1)   Make sure you come to the same accept and reject decisions as Table 7.1.

(7.2)   Using the QSS in Table 7.1 and assuming one is using the reduced inspection, what decision would you make if 3 defectives were found in the first sample? What sampling plan would you use to inspect the next lot?

7.2 Stationary and Transitive OC Curves

OC curves describe the protection provided by sampling plans. For single and double sampling plans, separate decisions are made for each lot inspected. Therefore, the protection that is provided does not depend on past lots. This is not the case for QSSs. Using a QSS, a lot is more likely to be rejected if previous lots where rejected and more likely to be released if previous lots where released. The protection provided by QSSs must be examined for both periods of constant quality and changing quality. Only Type-B OC curves apply to QSSs.

Stationary OC curves describe the protection provided by QSSs under periods of constant quality. Figure 7.2 shows the stationary OC curve of the QSS from the checker chart. The individual OC curves of the reduced and tightened sampling plans are also shown. The stationary OC curve is always between the OC curves of the individual sampling plans. Generally, the stationary OC curve approximates the OC curve of the reduced sampling plan at lower defect rates and approximates the OC curve of the tightened sampling plan at higher defect rates. However, in this example, the tightened and reduced OC curves are nearly identical at lower defect rates. As a result, the OC curve of the tightened sampling plan n=50 and a=1 closely matches the stationary OC curve. Appendix B provides formulas for calculating stationary OC curves.

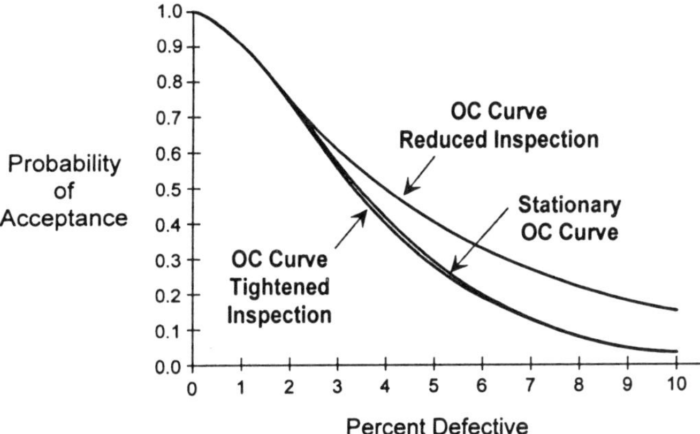

Figure 7.2: Stationary OC Curve of QSS
$n1_r=18$, $a1_r=0$, $r1_r=2$, $n2_r=80$, $a2_r=1$,
$n_t=50$, $a_t=1$ and $s_t=0$

Transitive OC curves describe the protection provided by QSSs under periods of changing quality. Assume the process has been running at p_{old} percent defective and suddenly jumps to p_{new}. The transitive OC curves give the probability of accepting the first lot following the change, the probability of accepting the second lot following the change, and so on. These are denoted $OC_T(p_{new}|p_{old},1)$, $OC_T(p_{new}|p_{old},2)$, and so on. Figure 7.3 shows the transitive OC curves of the QSS from the checker chart following a jump in quality from 0% defective. The first transitive OC curve, $OC_T(p_{new}|0\%,1)$ is the OC curve of the reduced sampling plan. The transitive OC curves quickly converge to the stationary OC curve. When jumping from a series of 100% defective lots, the first transitive OC curve, $OC_T(p_{new}|100\%,1)$, is the OC curve of the tightened sampling plan. Appendix B also provides formulas for calculating transitive OC curves.

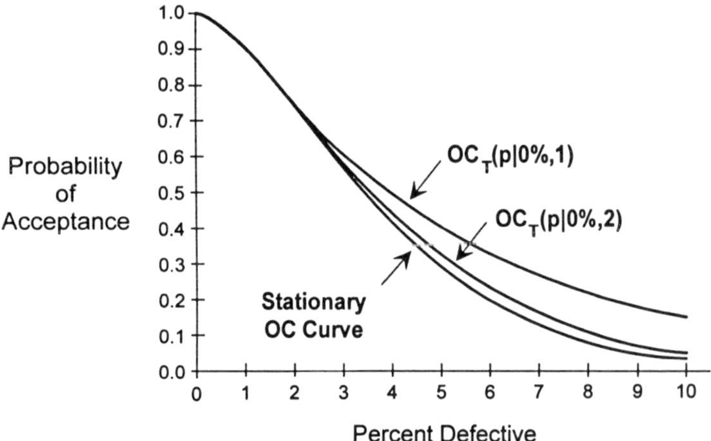

Figure 7.3: Transitive OC Curves of QSS
$n1_r=18$, $a1_r=0$, $r1_r=2$, $n2_r=80$, $a2_r=1$,
$n_t=50$, $a_t=1$ and $s_t=0$

Both stationary and transitive OC curves for QSSs can be obtained using the program QSS on the Sampling Programs diskette. The program's main menu is shown in Screen 7.1. Option 3 calculates OC curves. However, before selecting option 3, a QSS must be entered. Option 1 is used to enter QSSs. Screen 7.2 shows the use of option 1 to enter the QSS from the checker chart when defectives are tallied.

Screen 7.1: Main Menu (QSS)

```
****************************************************************************
*                           *                                              *
*                           *       Copyright (C) 1992 Taylor Enterprises  *
*       Program QSS         *         P.O. Box 820, Lake Villa, IL 60046   *
*                           *                (708) 356-1074                *
*                           *                                              *
****************************************************************************

                  Current Sampling Plan:  no plan selected



                       -----  MENU OPTIONS  -----
              (1)  Enter sampling plan
              (2)  Summary information
              (3)  OC curve
              (4)  AOQ curve
              (5)  ASN curve
              (6)  Select QSS based on AQL and LTPD
              (7)  Select QSS with specified reduced inspection


ENTER NUMBER OF OPTION OR "q" TO QUIT  -->
```

Screen 7.2: Entering QSS (QSS)

```
****************************************************************************
***                                                                     ***
***               QSS Option 1 - Enter Quick Switching System           ***
***                                                                     ***
****************************************************************************

   One must specify whether the reduced inspection uses a single or double
   sampling plan along with the value of the parameters of the QSS:

              Reduced Inspection                     Tightened Insection
         Single                  Double
      ------------------    ----------------------    --------------------
      nr = sample size      n1r = 1st sample size     nt = sample size
      ar = accept number    a1r = 1st accept number   at = accept number
                            r1r = 1st reject number    st = switch number
                            n2r = 2nd sample size
                            a2r = 2nd accept number

   One must also specify whether defects or defectives are tallied.

TALLY? (1=defectives,2=defects)           --> 1
REDUCED INSP.? ("S"=single, "D"=double)   --> d
ENTER FIRST SAMPLE SIZE REDUCED    (n1r)  --> 18
ENTER FIRST ACCEPT NUMBER REDUCED  (a1r)  --> 0
ENTER FIRST REJECT NUMBER REDUCED  (r1r)  --> 2
ENTER SECOND SAMPLE SIZE REDUCED   (n2r)  --> 80
ENTER SECOND ACCEPT NUMBER REDUC.  (a2r)  --> 1
ENTER SAMPLE SIZE TIGHTENED INSP.  (nt)   --> 50
ENTER ACCEPT NUMBER TIGHTENED INSP (at)   --> 1
ENTER SWITCH NUMBER TIGHTENED INSP (st)   --> 0
```

Once the QSS is entered, one can select option 3 to plot or tabulate
the OC curve. Screen 7.3 then appears requesting the type of OC
curve, whether a plot or table is desired, and information on scaling.
If a transitive OC curve is requested, the previous quality and

number of lots since the change are also required. Screen 7.3 shows
the input required to obtain a plot of the stationary OC curve using
the default scale. The resulting plot is displayed in Screen 7.4.
Screen 7.5 shows the same OC curve shown in the form of a table.

Screen 7.3: Entering OC Curve Information (QSS)

```
************************************************************************************
***                                                                            ***
***                      QSS Option 3 - OC Curve                               ***
***                                                                            ***
************************************************************************************

    This option plots or tabulates the OC curve.  You must specify which type
    of OC curve to use:

        (S) Stationary OC curve for periods of constant quality
        (T) Transitive OC curve for following a change in quality

    For transitive OC curves, the previous quality (POLD) and the number of lots
    since the change in quality must be specified.  One may specify the scale or
    let the program select the appropriate scale.

ENTER TYPE OF OC CURVE. ("S" or "T")     --> s
ENTER "p" FOR PLOT "t" FOR TABLE         --> p

    The default scale for percent defective (bottom axis) is from 0%
    to  10.0000000% defective.

USE DEFAULT SCALE? ("y" or "n")          --> y

    The default scale for probability of acceptance (left axis) is from 0 to 1.

USE DEFAULT SCALE? ("y" or "n")          --> y
```

Screen 7.4: Plot of Stationary OC Curve (QSS)

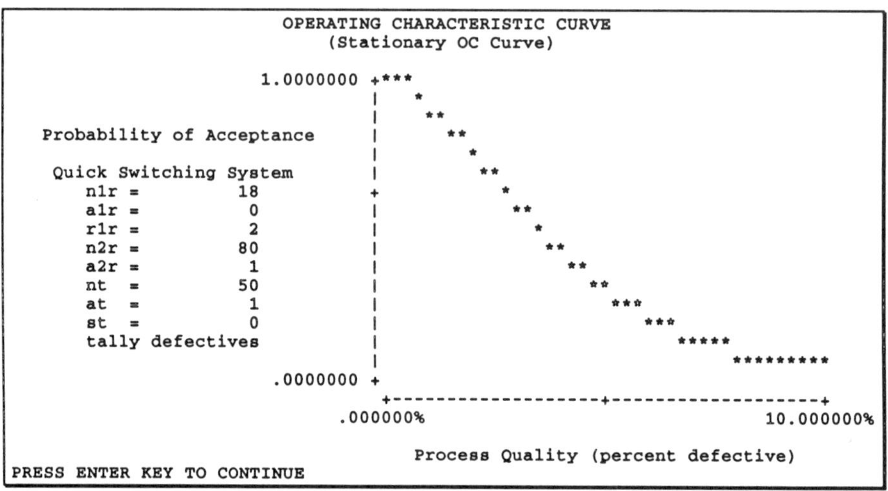

Screen 7.5: Table of Stationary OC Curve (QSS)

```
                    OPERATING CHARACTERISTIC CURVE
                         (Stationary OC Curve)

  Current Sampling Plan:  n1r =        18,   a1r =    0,   r1r =     2
                          n2r =        80,   a2r =    1
                          nt  =        50,   at  =    1,   st  =     0
                          tally defectives


          Percent Defective              Probability of Acceptance
       __________________             ____________________________

                .0000000                      1.00000000
               1.0000000                       .90354830
               2.0000000                       .74171470
               3.0000000                       .57338440
               4.0000000                       .41942660
               5.0000000                       .29359750
               6.0000000                       .19895040
               7.0000000                       .13161150
               8.0000000                       .08545877
               9.0000000                       .05464898
              10.0000000                       .03448730

 PRESS ENTER KEY TO CONTINUE
```

Exercises

(7.3)   Obtain a table and plot of the stationary OC curve of the QSS n_r=13, a_r=0, n_t=30, a_t=0 and s_t=0 when tallying defectives. What is the probability of acceptance when the process is running 1%, 4% and 8% defective?

(7.4)   Obtain a table and plot of the transitive OC curve of the QSS n_r=13, a_r=0, n_t=30, a_t=0 and s_t=0 for the first lot following a jump from 0% defective. Assume defectives are tallied. What is the probability of acceptance if the process jumps to 1%, 4% or 8% defective? Compare this transitive OC curve with the OC curve of the reduced inspection.

(7.5)   Obtain a table and plot of the transitive OC curve of the QSS n_r=13, a_r=0, n_t=30, a_t=0 and s_t=0 for the first lot following a jump from 100% defective. Assume defectives are tallied. What is the probability of acceptance if the process jumps to 1%, 4% or 8% defective? Compare this transitive OC curve with the OC curve of the tightened inspection.

(7.6)   Obtain a table and plot of the stationary OC curve of the QSS n_r=13.0, a_r=0, n_t=30.0, a_t=0 and s_t=0 when one is inspecting for defects from a continuous lot. What is the probability of acceptance when the process is averaging 0.01, 0.04 and 0.08 defects per unit?

7.3   Summarizing Protection

The AQL, IQ, LTPD and AOQL summarize the protection provided by single and double sampling plans. They can also be calculated for QSSs based on the stationary OC curve. However, these summary statistics do not describe the protection provided by a QSSs during periods of changing quality. For QSSs four additional summary statistics are required.

Following a change in quality, a QSS has an increased chance of making an error. Following a series of good quality lots, the first bad lot has an increased chance of acceptance. Likewise, following a series of bad quality lots, the first good lot has an increased chance of rejection. Of interest is the maximum chance of making these two errors. Figure 7.4 shows how to determine the maximum chance of making an error at the AQL and LTPD.

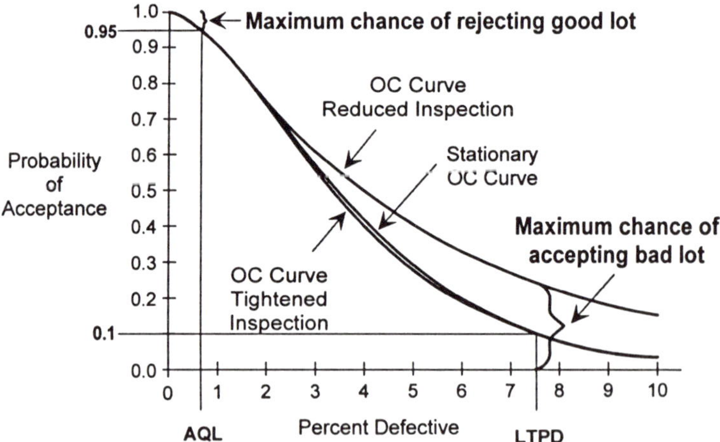

Figure 7.4: Maximum Chance of Two Types of Errors for QSS
$n1_r$=18, $a1_r$=0, $r1_r$=2, $n2_r$=80, $a2_r$=1,
n_t=50, a_t=1 and s_t=0

For each subsequent lot the increased chance of making an error is reduced as the transitive OC curves converge to stationary OC curve. Also of interest is the rate of convergence. Figure 7.5 shows how the rate of convergence is determined. The increased risk is reduced by 72% for the second lot following the change. It will continue to decrease at the same rate for subsequent lots.

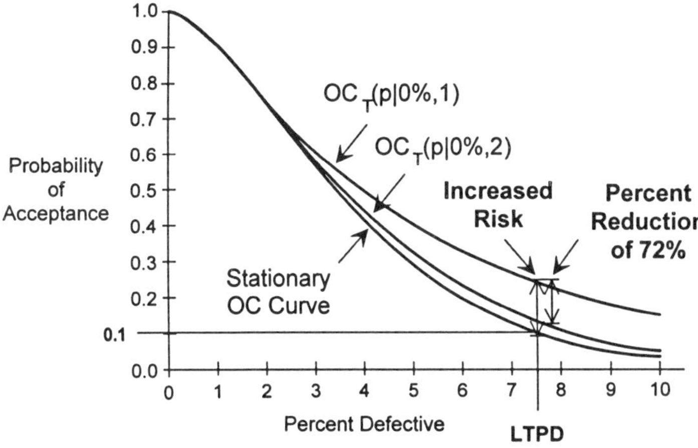

Figure 7.5: Rate of Convergence at LTPD for QSS
$n1_r$=18, $a1_r$=0, $r1_r$=2, $n2_r$=80, $a2_r$=1,
n_t=50, a_t=1 and s_t=0

Option 2 of program QSS can be used to obtain the AQL, IQ, LTPD, AOQL, maximum chances of errors, and rates of convergence for a QSS. Screen 7.6 shows the summary information for the QSS from the checker chart when defectives are tallied.

Screen 7.6: Summary Information (QSS)

```
********************************************************************************
***                  QSS Option 2 - Summary Statistics                     ***
********************************************************************************

     Current Sampling Plan:   n1r =          18,    a1r =      0,    r1r =     2
                              n2r =          80,    a2r =      1
                              nt  =          50,    at  =      1,    st  =     0
                              tally defectives

                          STATIONARY OC CURVE
               Percentile        Symbol      Percent Defective
               ----------        ------      -----------------
                  95.0%           AQL              .6644519
                  90.0%                           1.0238080
                  50.0%           IQ              3.4580510
                  10.0%           LTPD            7.6405020
                   5.0%                           9.1952950
                                  AOQL            1.7320310

        Maximum chance of rejection at AQL     =    .0504349
        Rate of convergence at AQL             =    71.65294%
        Maximum chance of acceptance at LTPD   =    .2397654
        Rate of convergence at LTPD            =    76.02346%

PRESS ENTER KEY TO CONTINUE
```

The QSS from the checker chart has an AQL of 0.66% and LTPD of 7.64%. These are very close to those of the double sampling plan n1=32, a1=0, r1=2, n2=32 and a2=1 (AQL=0.65%, LTPD=7.50%) and the single sampling plan n=50 and a=1 (AQL=0.72%, LTPD=7.56%). However, during periods of changing quality, the QSS has increased chances of making errors. They are:

Maximum chance of rejection at AQL	= 0.0504
Rate of convergence at AQL	= 71.7%
Maximum chance of acceptance at LTPD	= 0.240
Rate of convergence at LTPD	= 76.0%

The chance of this QSS accepting a bad lot increases to 0.24 for the first bad lot following a change and then rapidly converges back to 0.1. The increase chance of acceptance is reduced by 72% for each subsequent lot. There is no increased risk for good lots. The increased risk of this QSS under changing quality is small.

Exercises

(7.7)   What are the AQL, IQ, LTPD, AOQL, maximum chances of errors and rates of convergence of the QSS n_r=13, a_r=0, n_t=30, a_t=0 and s_t=0 when tallying defectives?

(7.8)   What are the AQL, IQ, LTPD, AOQL, maximum chances of errors and rates of convergence of the QSS n_r=13, a_r=0, n_t=30, a_t=0 and s_t=0 when tallying defects from a continuous lot?

7.4   AOQ Curves

Option 4 of QSS can be used to obtain plots and tables of the AOQ curve. The only assumption made is that rejected lots are 100% inspected. One must specify the efficiency of the 100% inspection, the lot size, and the method of handling defectives or defects. Values of the AOQ are calculated using the formulas given in Chapter 2 with the ASN of the QSS substituted for the sample size of the single sampling plan.

Assume the QSS from the checker chart has been entered. Screen 7.7 shows the input required to obtain a plot of the AOQ curve when the 100% inspection is 100% efficient, the lot size is large and defectives are replaced. These are the conditions assumed under option 2 when calculating the AOQL. Screen 7.8 shows the resulting plot.

Screen 7.7: Entering AOQ Curve Information (QSS)

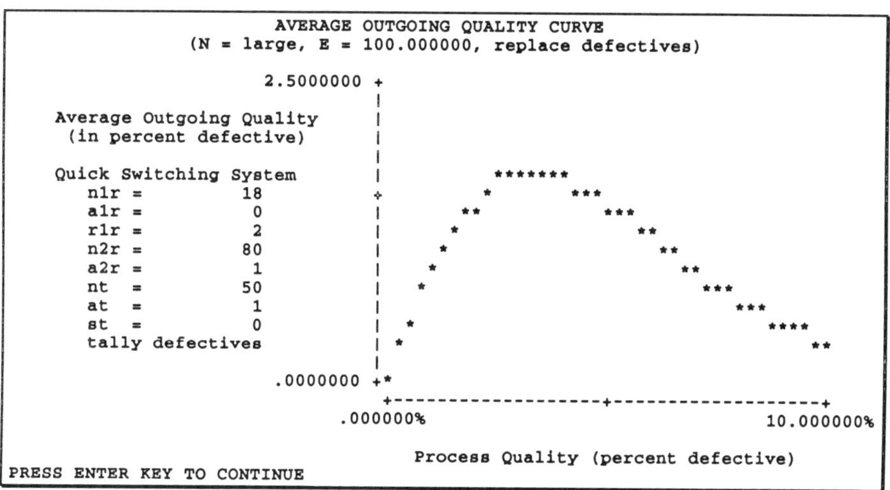

```
**********************************************************************************
***                                                                          ***
***                      QSS Option 4 - AOQ Curve                            ***
***                                                                          ***
**********************************************************************************

   This option plots or tabulates the AOQ curve.  It is assumed that
   rejected lots are 100% inspected.  One must enter the efficiency of
   the 100% inspection, the lot size, and whether defects and
   defectives found are either repaired, replaced or discarded.
   One can specify the scale or let the program select the appropriate scale.

ENTER EFFICIENCY OF 100% INSPECTION      --> 100
ENTER LOT SIZE (0 = Large)               --> 0
DEFECTS? (1=repair,2=replace,3=discard) --> 2
ENTER "p" FOR PLOT "t" FOR TABLE         --> p

   The default scale for percent defective (bottom axis) is from 0%
   to  10.0000000% defective.

USE DEFAULT SCALE? ("y" or "n")          --> y

   The default scale for average outgoing quality (left axis) is from 0%
   to   2.5000000% defective.

USE DEFAULT SCALE? ("y" or "n")          --> y
```

Screen 7.8: Plot of AOQ Curve (QSS)

```
                        AVERAGE OUTGOING QUALITY CURVE
              (N = large, E = 100.000000, replace defectives)

                    2.5000000 +
                              |
    Average Outgoing Quality  |
      (in percent defective)  |
                              |
    Quick Switching System    |                    *******
        n1r =          18     +              *           ***
        a1r =           0     |            **               ***
        r1r =           2     |           *                    **
        n2r =          80     |          *                        **
        a2r =           1     |         *                           **
        nt  =          50     |        *                              ***
        at  =           1     |      *                                   ***
        st  =           0     |     *                                      ****
        tally defectives      |    *                                          **
                              |
                    .0000000 +*
                              +------------------------+------------------+
                          .000000%                                  10.000000%

                          Process Quality (percent defective)

PRESS ENTER KEY TO CONTINUE
```

Exercise

(7.9)   Obtain a plot and table of the AOQ curve for the QSS $n1_r$=18,
$a1_r$=0, $r1_r$=2, $n2_r$=80, $a2_r$=1, n_t=50, a_t=1 and s_t=0. Assume
defectives are replaced, the lot size is large, and the 100%
inspection is 90% efficient. What is the AOQ at 1%, 4% and
8% defective?

7.5   ASN Curves

When using QSSs, the number of units inspected varies from lot to lot. The number of units inspected depends on whether the tightened or reduced sampling plan is used. It also depends on whether a second sample is required while using the reduced sampling plan. The probability of using the reduced sampling plan and the probability of taking a second sample depend on the process quality. The ASN curve is a plot of the average number of units inspected versus the process quality.

The program QSS can be used to obtain plots and tables of ASN curves for QSSs. The sampling plan must be entered first. ASN curves are then obtained by selecting option 5 from the main menu. Screen 7.9 then appears asking whether a plot or table is desired and requesting information on scaling. Assume that the QSS $n1_r=18$, $a1_r=0$, $r1_r=2$, $n2_r=80$, $a2_r=1$, $n_t=50$, $a_t=1$ and $s_t=0$ has already been entered where defectives are tallied. Screen 7.9 shows the input required to obtain a plot of the ASN curve using the default scaling. The resulting plot is displayed in Screen 7.10. Option 5 assumes that no curtailing is performed.

Screen 7.9: Entering ASN Curve Information (QSS)

```
*************************************************************************************
***                                                                             ***
***                    QSS Option 5 - ASN Curve                                  ***
***                                                                             ***
*************************************************************************************

    This option plots or tabulates the ASN curve.  One may specify the
    scale or let the program select the appropriate scale.

ENTER "p" FOR PLOT "t" FOR TABLE            --> p

    The default scale for percent defective (bottom axis) is from 0%
    to   10.0000000% defective.

USE DEFAULT SCALE? ("y" or "n")             --> y

    The default scale for average sample number (left axis) is from 0
    to        98.00000.

USE DEFAULT SCALE? ("y" or "n")             --> y
```

Screen 7.10: Plot of ASN Curve (QSS)

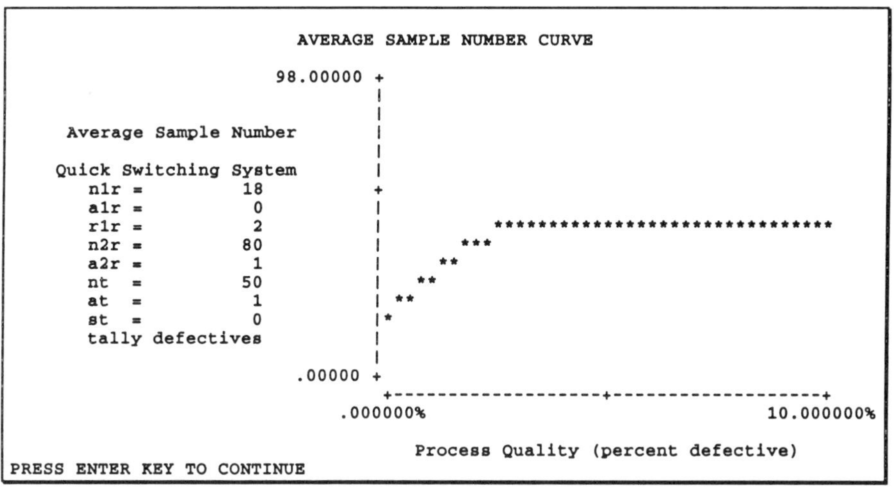

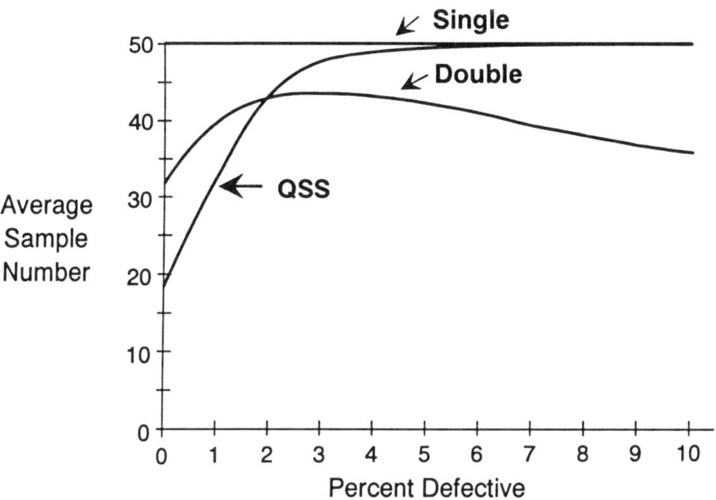

Figure 7.6: Average Sample Number Curves for
QSS $n1_r$=18, $a1_r$=0, $r1_r$=2, $n2_r$=80, $a2_r$=1, n_t=50, a_t=1, s_t=0
Single Sampling Plan n=50, a=1 and
Double Sampling Plan n1=32, a1=0, r1=2, n2=32, a2=2

QSSs can be used in place of single and double sampling plans to reduce the amount of inspection. Figure 7.6 shows the ASN curve of the QSS from the checker chart along with the ASN curves of the matching single and double sampling plans. At lower defect rates, this QSS reduces the number of units inspected by 26% to 44% compared to the double sampling plan; and by 44% to 64% compared to the single sampling plan.

QSSs do not, however, provide equivalent protection to single and double sampling plans. Matching OC curves only ensures that they provide equivalent protection during periods of constant quality. During periods of changing quality, the QSS has increased risks of accepting bad lots and of rejecting good lots. For the QSS $n1_r=18$, $a1_r=0$, $r1_r=2$, $n2_r=80$, $a2_r=1$, $n_t=50$, $a_t=1$ and $s_t=0$, the only risk that increases is the risk of accepting a bad lot. This risk doubles on the first lot following a change in quality. The increased risk largely disappears by the second lot. The increased risk is small compared to the resulting reduction in the number of units inspected.

In general, it is possible to select QSSs with moderate increases in the chances of making errors during periods of changing quality that offer sizable reductions in the number of samples selected. It is also possible to make even further reductions in the sample size by allowing increased chances of making errors. In the next section, it is shown how to control the increased chances of errors and the rates of convergence when selecting QSSs.

QSSs also have other disadvantages. The inspection procedure is more complicated. Further, the maximum number of units inspected from a lot is greater: 98 versus 64 and 50. This may require setting more units aside. Finally, like double sampling, two samples are required from some lots. In some circumstances these factors may result in the selection of the single or double sampling plan instead. However, most times these complications are minor compared to the reduction in sample size. A well-designed checker chart can alleviate much of the added complexity.

Exercise

(7.10)   Obtain a plot and table of the ASN curve of the QSS $n_r=13$, $a_r=0$, $n_t=30$, $a_t=0$ and $s_t=0$ when tallying defective units. What is the ASN at 1%, 4% and 8% defective?

7.6   Selecting QSSs Based on AQL and LTPD

QSSs selected on the basis of their LTPD are used to protect against the release of highly defective lots. The LTPD represents a level of defects to guard against. Specifying an AQL protects against the rejection of good lots. For QSSs, the maximum chances of making errors and the rates of convergence should also be specified. To select a QSS that can be safely used in place of single and double sampling plans, the following restrictions are used:

Maximum chance of rejection at AQL	$\leq$ 0.15
Rate of convergence at AQL	$\geq$ 50%
Maximum chance of acceptance at LTPD	$\leq$ 0.3
Rate of convergence at LTPD	$\geq$ 70%

Of the QSSs having the specified AQL and LTPD and meeting these restrictions, the one minimizing the amount of inspection is selected.

The above restrictions triple the chances of making errors for the first lot following a change. Following a jump to the AQL, the chance of rejection can triple from 0.05 to 0.15. The rate of convergence at the AQL ensures the chance of rejection decreases to 0.1 for the second lot, 0.75 for the third lot and so on. Following a jump to the LTPD, the chance of acceptance can triple from 0.1 to 0.3. The rate of convergence at the LTPD ensures the chance of acceptance decreases to 0.16 for the second lot, 0.12 for the third lot and so on. The increased chance of accepting a bad lot largely disappears by the second lot.

The program QSS can be used to select QSSs based on their AQL, LTPD, maximum chances of making errors and the rates of convergence. To select a QSS, select option 6 from the main menu. The program then asks for the AQL and LTPD, whether defectives or defects are tallied and whether to use a single or double sampling plan for reduced inspection. One must also specify the process quality p_0 at which to minimize the ASN. The QSS is selected that minimizes the ASN at p_0 from those satisfying the above restrictions and the conditions:

Actual AQL	$\geq$	Specified AQL
Actual LTPD	$\leq$	Specified LTPD

The definitions of the AQL (95%) and LTPD (10%) as well as the values used for the four restrictions can be changed.

Suppose one wishes to select the QSS with an AQL of 0.65% and LTPD of 8% that minimizes the ASN at 0.65%. A double sampling plan is desired for reduced inspection. Assume defectives are tallied. Screen 7.11 shows the required input. The program displays the selected QSS at the top of the main menu (Screen 7.12). The selected QSS is $n1_r=15$, $a1_r=0$, $r1_r=2$, $n2_r=40$, $a2_r=1$, $n_t=48$, $a_t=1$ and $s_t=0$. The procedure used to select QSSs is explained in Taylor (1992b, 1992c)

Screen 7.11: Selecting QSS Based on AQL and LTPD (QSS)

```
***********************************************************************************
***                                                                           ***
***               QSS Option 6 - Select QSS Based on AQL and LTPD             ***
***                                                                           ***
***********************************************************************************

    This option selects a QSS minimizing the ASN at a specified point P0
    from among those satisfying the following conditions:

        Actual AQL is greater than or equal to the specified AQL
        Maximum chance of rejection at AQL is 0.15
        Rate of convergence at AQL is at least 50%
        Actual LTPD is less than or equal to the specified LTPD
        Maximum chance of acceptance at LTPD is 0.3
        Rate of convergence at LTPD is at least 70%

    One must specify whether the reduced inspection should be a single or
    double sampling plan.  The standard definitions of AQL (95% chance of
    acceptance), LTPD (10% chance of acceptance) and the values given above
    are used unless indicated otherwise.

TALLY? (1=defectives,2=defects)              --> 1
REDUCED INSP.? ("S"=single, "D"=double)  --> d
ENTER AQL   AS PERCENT DEFECTIVE             --> 0.65
ENTER LTPD AS PERCENT DEFECTIVE              --> 8.0
ENTER P0    AS PERCENT DEFECTIVE             --> 0.65
USE DEFAULT DEFINITIONS? ("y" or "n")        --> y
```

Screen 7.12: Main Menu Showing Selected QSS (QSS)

```
***********************************************************************************
*                         *                                                     *
*                         *     Copyright (C) 1992 Taylor Enterprises           *
*     Program QSS         *        P.O. Box 820, Lake Villa, IL 60046           *
*                         *               (708) 356-1074                        *
*                         *                                                     *
***********************************************************************************

      Current Sampling Plan:  n1r =        15,    a1r =     0,   r1r =    2
                              n2r =        40,    a2r =     1
                              nt  =        48,    at  =     1,   st  =    0
                              tally defectives

                          ----- MENU OPTIONS -----

                (1) Enter sampling plan
                (2) Summary information
                (3) OC curve
                (4) AOQ curve
                (5) ASN curve
                (6) Select QSS based on AQL and LTPD
                (7) Select QSS with specified reduced inspection

ENTER NUMBER OF OPTION OR "q" TO QUIT  -->
```

An alternative approach to selecting QSSs with a specified AQL and LTPD is to use Table 7.2. This table assumes that defectives are tallied. Table 7.2 covers the same range of AQLs and LTPDs as Table 3.3 for single sampling plans and Table 6.3 for double sampling plans. The reduced inspection is given first, followed by the tightened inspection and the switch number for switching from tightened to reduced. Except for the first column, the reduced inspections are double sampling plans. The QSSs in Table 7.2 minimize the ASN at the AQL. They also satisfy the four restrictions given on page 159.

The first column of Table 7.2 corresponds to single sampling plans with zero acceptance numbers. In this case, there is no QSS providing equivalent protection at a reduced cost. However, QSSs can be found providing improved protection at the same cost. These QSSs are given in the first column. The column label has been changed from 45 to 24 to indicate the improved protection.

When the program QSS was used to select a QSS with an AQL of 0.65% and LTPD of 8%, the resulting QSS was $n1_r$=15, $a1_r$=0, $r1_r$=2, $n2_r$=40, $a2_r$=1, n_t=48, a_t=1 and s_t=0. The QSS in Table 7.2 closest to these values is $n1_r$=18, $a1_r$=0, $r1_r$=2, $n2_r$=80, $a2_r$=1, n_t=50, a_t=1 and s_t=0. This QSS is found in the row labeled "0.65%" and the column labeled "11". All three sample sizes are increased. Why the difference? The reason is that the program QSS only requires that the actual AQL be greater than or equal to the specified AQL. The QSS selected by the program QSS has an actual AQL of 0.947%. This is significantly larger than the specified AQL. Table 7.2, on the other hand, requires that the actual AQL be approximately equal to the specified AQL. Table 7.2 has an actual AQL of 0.664%. This is much closer to the specified AQL. This added requirement can drive up the sample sizes.

Table 7.2 can also be used to select QSSs when defects are tallied so long as the defect rate is low ($\leq$0.1). Otherwise, the program QSS should be used. When using QSS for defects, the AQL, LTPD and p_0 must be given in terms of the average number of defects per unit. Further, QSS will ask if the lot is continuous or consists of units of product. For continuous lots, non integer sample sizes are possible. One must specify the smallest sample size increment: n_{inc}. QSS will then restrict the sample sizes to: n_{inc}, $2n_{inc}$, $3n_{inc}$, $\cdots$. Setting n_{inc} equal to one ensures integer sample sizes.

Table 7.2: QSSs by AQL and LTPD for Defectives

AQL	Approximate Ratio of $LTPD/AQL$								
	24 †	11	6.5	5	4	3.2	2.8	2.3	2
10%	-	2/(0,2) 2/(1,2) 4/(1,2) 0 AQL = 10.6 LTPD = 68.0 AOQL = 20.3	2/(0,2) 8/(2,3) 7/(1,2) 0 AQL = 9.82 LTPD = 45.8 AOQL = 13.8	3/(0,2) 10/(3,4) 12/(2,3) 1 AQL = 10.1 LTPD = 39.1 AOQL = 13.6	4/(0,3) 10/(3,4) 18/(3,4) 2 AQL = 10.7 LTPD = 34.1 AOQL = 13.1	10/(1,4) 10/(4,5) 25/(4,5) 2 AQL = 10.2 LTPD = 29.6 AOQL = 11.7	12/(1,5) 30/(7,8) 36/(5,6) 3 AQL = 9.93 LTPD = 24.4 AOQL = 10.4	18/(2,7) 50/(10,11) 60/(8,9) 6 AQL = 9.82 LTPD = 20.9 AOQL = 10.0	28/(3,10) 60/(13,14) 84/(11,12) 8 AQL = 10.0 LTPD = 19.2 AOQL = 9.94
6.5%	-	2/(0,2) 7/(1,2) 6/(1,2) 0 AQL = 6.49 LTPD = 51.3 AOQL = 14.0	3/(0,2) 12/(2,3) 10/(1,2) 0 AQL = 6.49 LTPD = 34.1 AOQL = 9.61	4/(0,2) 18/(3,4) 16/(2,3) 1 AQL = 6.59 LTPD = 30.3 AOQL = 9.49	6/(0,3) 17/(3,4) 28/(3,4) 2 AQL = 6.53 LTPD = 22.8 AOQL = 8.25	13/(1,4) 30/(5,6) 40/(4,5) 2 AQL = 6.63 LTPD = 19.1 AOQL = 7.41	20/(1,5) 42/(7,8) 65/(6,7) 4 AQL = 6.54 LTPD = 15.8 AOQL = 6.91	30/(2,7) 80/(11,12) 90/(8,9) 6 AQL = 6.58 LTPD = 14.1 AOQL = 6.67	45/(3,10) 100/(14,15) 135/(11,12) 8 AQL = 6.54 LTPD = 12.1 AOQL = 6.43
4.0%	-	3/(0,2) 13/(1,2) 9/(1,2) 0 AQL = 4.00 LTPD = 37.1 AOQL = 9.43	5/(0,2) 18/(2,3) 15/(1,2) 0 AQL = 4.08 LTPD = 23.8 AOQL = 6.32	9/(0,3) 28/(3,4) 34/(2,3) 1 AQL = 3.97 LTPD = 15.2 AOQL = 5.00	10/(0,4) 25/(3,4) 45/(3,4) 2 AQL = 4.21 LTPD = 14.6 AOQL = 5.26	20/(1,4) 50/(5,6) 50/(3,4) 2 AQL = 4.10 LTPD = 13.1 AOQL = 4.74	28/(1,5) 75/(7,8) 100/(6,7) 4 AQL = 4.07 LTPD = 10.4 AOQL = 4.37	50/(2,7) 110/(10,11) 150/(8,9) 6 AQL = 3.97 LTPD = 8.60 AOQL = 4.05	70/(3,10) 165/(14,15) 210/(11,12) 8 AQL = 4.04 LTPD = 7.84 AOQL = 3.99
2.5%	2/(0,1) 5/(0,1) 0 AQL = 2.36 LTPD = 40.9 AOQL = 9.57	4/(0,2) 25/(1,2) 14/(1,2) 0 AQL = 2.59 LTPD = 25.4 AOQL = 6.31	8/(0,2) 28/(2,3) 24/(1,2) 0 AQL = 2.56 LTPD = 15.5 AOQL = 4.01	11/(0,3) 26/(2,3) 42/(2,3) 1 AQL = 2.56 LTPD = 12.4 AOQL = 3.78	18/(0,4) 40/(3,4) 70/(3,4) 2 AQL = 2.51 LTPD = 9.47 AOQL = 3.24	35/(1,5) 50/(4,5) 80/(3,4) 2 AQL = 2.48 LTPD = 8.31 AOQL = 2.95	45/(1,6) 120/(7,8) 170/(6,7) 4 AQL = 2.52 LTPD = 6.18 AOQL = 2.67	75/(2,7) 180/(10,11) 230/(8,9) 6 AQL = 2.52 LTPD = 5.64 AOQL = 2.59	110/(3,9) 265/(14,15) 340/(11,12) 8 AQL = 2.52 LTPD = 4.87 AOQL = 2.49
1.5%	3/(0,1) 7/(0,1) 0 AQL = 1.59 LTPD = 31.0 AOQL = 6.84	7/(0,2) 40/(1,2) 20/(1,2) 0 AQL = 1.55 LTPD = 18.3 AOQL = 4.26	13/(0,2) 40/(2,3) 40/(1,2) 0 AQL = 1.63 LTPD = 9.54 AOQL = 2.50	16/(0,3) 50/(2,3) 65/(2,3) 1 AQL = 1.52 LTPD = 8.14 AOQL = 2.37	22/(0,3) 75/(3,4) 85/(2,3) 1 AQL = 1.53 LTPD = 6.27 AOQL = 1.98	50/(1,4) 90/(4,5) 130/(3,4) 2 AQL = 1.54 LTPD = 5.19 AOQL = 1.85	75/(1,5) 150/(6,7) 230/(5,6) 3 AQL = 1.49 LTPD = 4.03 AOQL = 1.62	110/(2,7) 270/(9,10) 340/(7,8) 5 AQL = 1.51 LTPD = 3.48 AOQL = 1.56	170/(3,10) 410/(13,14) 500/(10,11) 7 AQL = 1.51 LTPD = 3.08 AOQL = 1.51
1.0%	5/(0,1) 11/(0,1) 0 AQL = 0.965 LTPD = 20.8 AOQL = 4.37	10/(0,2) 70/(1,2) 32/(1,2) 0 AQL = 1.00 LTPD = 11.8 AOQL = 2.75	20/(0,2) 60/(2,3) 60/(1,2) 0 AQL = 1.07 LTPD = 6.44 AOQL = 1.66	28/(0,3) 60/(2,3) 100/(2,3) 1 AQL = 1.05 LTPD = 5.33 AOQL = 1.58	42/(0,4) 100/(3,4) 170/(3,4) 2 AQL = 1.03 LTPD = 3.97 AOQL = 1.34	90/(1,5) 110/(4,5) 200/(3,4) 2 AQL = 1.02 LTPD = 3.38 AOQL = 1.21	110/(1,7) 300/(7,8) 400/(6,7) 4 AQL = 1.02 LTPD = 2.64 AOQL = 1.10	190/(2,9) 440/(10,11) 600/(8,9) 6 AQL = 1.01 LTPD = 2.18 AOQL = 1.03	270/(3,10) 670/(14,15) 840/(11,12) 8 AQL = 1.01 LTPD = 1.98 AOQL = 1.00

† Provide better protection at the same cost as single sampling plans in Table 3.3.

Table 7.2: (continued)

AQL	Approximate Ratio of $LTPD/AQL$								
	24 †	11	6.5	5	4	3.2	2.8	2.3	2
0.65%	8/(0,1) 18/(0,1) 0 AQL = 0.603 LTPD = 13.4 AOQL = 2.74	18/(0,2) 80/(1,2) 50/(1,2) 0 AQL = 0.664 LTPD = 7.64 AOQL = 1.73	32/(0,2) 94/(2,3) 94/(1,2) 0 AQL = 0.676 LTPD = 4.14 AOQL = 1.06	50/(0,3) 80/(2,3) 165/(2,3) 1 AQL = 0.676 LTPD = 3.26 AOQL = 0.983	75/(0,4) 140/(3,4) 270/(3,4) 2 AQL = 0.660 LTPD = 2.51 AOQL = 0.850	125/(1,4) 300/(5,6) 400/(4,5) 2 AQL = 0.660 LTPD = 2.01 AOQL = 0.748	180/(1,7) 450/(7,8) 650/(6,7) 4 AQL = 0.653 LTPD = 1.63 AOQL = 0.694	290/(2,7) 800/(11,12) 1000/(9,10) 6 AQL = 0.656 LTPD = 1.42 AOQL = 0.665	440/(3,10) 1000/(14,15) 1450/(12,13) 8 AQL = 0.647 LTPD = 1.23 AOQL = 0.635
0.4%	13/(0,1) 30/(0,1) 0 AQL = 0.370 LTPD = 8.27 AOQL = 1.68	26/(0,2) 160/(1,2) 80/(1,2) 0 AQL = 0.406 LTPD = 4.85 AOQL = 1.10	52/(0,2) 150/(2,3) 150/(1,2) 0 AQL = 0.419 LTPD = 2.61 AOQL = 0.660	70/(0,3) 150/(2,3) 260/(2,3) 1 AQL = 0.419 LTPD = 2.08 AOQL = 0.619	110/(0,4) 240/(3,4) 425(3,4) 2 AQL = 0.412 LTPD = 1.60 AOQL = 0.536	200/(1,4) 500/(5,6) 600/(4,5) 2 AQL = 0.410 LTPD = 1.34 AOQL = 0.478	280/(1,7) 740/(7,8) 1000/(6,7) 4 AQL = 0.409 LTPD = 1.06 AOQL = 0.439	475/(2,9) 1100/(10,11) 1500/(8,9) 6 AQL = 0.402 LTPD = 0.876 AOQL = 0.410	-
0.25%	20/(0,1) 44/(0,1) 0 AQL = 0.242 LTPD = 5.67 AOQL = 1.13	40/(0,2) 250/(1,2) 130/(1,2) 0 AQL = 0.261 LTPD = 3.01 AOQL = 0.688	80/(0,2) 240/(2,3) 240/(1,2) 0 AQL = 0.267 LTPD = 1.64 AOQL = 0.417	110/(0,3) 250/(2,3) 400/(2,3) 1 AQL = 0.260 LTPD = 1.35 AOQL = 0.394	170/(0,4) 400/(3,4) 700/(3,4) 2 AQL = 0.254 LTPD = 0.974 AOQL = 0.329	350/(1,5) 450/(4,5) 800/(3,4) 2 AQL = 0.256 LTPD = 0.851 AOQL = 0.303	450/(1,6) 1200/(7,8) 1700/(6,7) 4 AQL = 2.51 LTPD = 0.626 AOQL = 0.267	-	-
0.15%	32/(0,1) 70/(0,1) 0 AQL = 0.151 LTPD = 3.60 AOQL = 0.713	65/(0,2) 400/(1,2) 210/(1,2) 0 AQL = 0.162 LTPD = 1.87 AOQL = 0.426	130/(0,2) 380/(2,3) 380/(1,2) 0 AQL = 0.166 LTPD = 1.04 AOQL = 0.262	170/(0,3) 440/(2,3) 650/(2,3) 1 AQL = 0.158 LTPD = 0.834 AOQL = 0.241	230/(0,3) 700/(3,4) 850/(2,3) 1 AQL = 0.155 LTPD = 0.639 AOQL = 0.200	520/(1,4) 850/(4,5) 1300/(3,4) 2 AQL = 0.154 LTPD = 0.525 AOQL = 0.185	-	-	-
0.1%	50/(0,1) 110/(0,1) 0 AQL = 0.0969 LTPD = 2.31 AOQL = 0.456	100/(0,2) 700/(1,2) 320/(1,2) 0 AQL = 0.100 LTPD = 1.23 AOQL = 0.278	200/(0,2) 600/(2,3) 600/(1,2) 0 AQL = 0.106 LTPD = 0.658 AOQL = 0.167	280/(0,3) 600/(2,3) 1000/(2,3) 1 AQL = 0.105 LTPD = 0.542 AOQL = 0.158	425/(0,4) 1000/(3,4) 1700/(3,4) 2 AQL = 0.102 LTPD = 0.401 AOQL = 0.134	-	-	-	-
0.065%	80/(0,1) 175/(0,1) 0 AQL = 0.0606 LTPD = 1.46 AOQL = 0.287	180/(0,2) 800/(1,2) 510/(1,2) 0 AQL = 0.0664 LTPD = 0.770 AOQL = 0.172	325/(0,2) 950/(2,3) 950/(1,2) 0 AQL = 0.0665 LTPD = 0.416 AOQL = 0.105	500/(0,3) 800/(2,3) 1600/(2,3) 1 AQL = 0.0677 LTPD = 0.339 AOQL = 0.100	-	-	-	-	-

† Provide better protection at the same cost as single sampling plans in Table 3.3.

Table 7.2: (continued)

AQL	Approximate Ratio of $LTPD/AQL$								
	24 †	**11**	**6.5**	**5**	**4**	**3.2**	**2.8**	**2.3**	**2**
0.04%	125/(0,1) 275/(0,1) 0 AQL = 0.0388 LTPD = 0.930 AOQL = 0.183	260/(0,2) 1600/(1,2) 800/(1,2) 0 AQL = 0.0406 LTPD = 0.493 AOQL = 0.111	500/(0,2) 1500/(2,3) 1500/(1,2) 0 AQL = 0.0427 LTPD = 0.264 AOQL = 0.0668	.	.	.	.	.	.
0.025%	200/(0,1) 440/(0,1) 0 AQL = 0.0242 LTPD = 0.582 AOQL = 0.114	400/(0,2) 2500/(1,2) 1300/(1,2) 0 AQL = 0.0261 LTPD = 0.304 AOQL = 0.0691	.	.	.	.	.	.	.
0.015%	315/(0,1) 700/(0,1) 0 AQL = 0.0154 LTPD = 0.367 AOQL = 0.0723	650/(0,2) 4000/(1,2) 2100/(1,2) 0 AQL = 0.0162 LTPD = 0.188 AOQL = 0.0427	.	.	.	.	.	.	.
0.01	500/(0,1) 1100/(0,1) 0 AQL = 0.00964 LTPD = 0.233 AOQL = 0.0458	.	.	.	.	.	.	.	.
0.0065%	800/(0,1) 1750/(0,1) 0 AQL = 0.00606 LTPD = 0.147 AOQL = 0.0287	.	.	.	.	.	.	.	.
0.004%	1250/(0,1) 2750/(0,1) 0 AQL = 0.00388 LTPD = 0.0934 AOQL = 0.0183	.	.	.	.	.	.	.	.
0.0025%	2000/(0,1) 4400/(0,1) 0 AQL = 0.00242 LTPD = 0.0584 AOQL = 0.0115	.	.	.	.	.	.	.	.

† Provide better protection at the same cost as single sampling plans in Table 3.3.

Exercises

(7.11) Select the QSS with a double sampling plan for reduced inspection and with an AQL of 0.4% and LTPD of 5% that minimizes the ASN at 0.4%. Assume defectives are tallied. Use both the program and table.

(7.12) Select the QSS with a double sampling plan for reduced inspection and with an AQL of 0.4% and LTPD of 5% that minimizes the ASN at 0.01%. Assume defectives are tallied.

(7.13) Select the QSS with a double sampling plan for reduced inspection and with an AQL of 10.0 defects per unit and LTPD of 50.0 defects per unit that minimizes the ASN at 10.0 defects per unit. Assume lots are continuous and use a smallest sample size increment of 0.01.

7.7   Using QSSs to Improve Protection

In the previous section it was shown how to select QSSs offering nearly the same protection as single and double sampling plans. These QSSs significantly reduce the number of units inspected. An alternative use of QSSs is to offer increased protection at the same cost. For example, suppose the single sampling plan n=13 and a=0 is used. This sampling plan has an AQL of 0.394% and LTPD of 16.2%. This sampling plan is indexed under an AQL of 0.4% and a ratio of 45. Table 7.2 gives a QSS that offers improved protection at the same cost. This is the QSS n_r=13, a_r=0, n_t=30, a_t=0 and s_t=0. It has an AQL of 0.370% and LTPD of 8.27%. While the AQL is changed very little, a dramatic improvement is made in the LTPD. The ASN for this QSS is 13 at 0% defective, 13.5 at 0.2% defective, and 13.9 at 0.4% defective. The resulting increase in the number of samples ranges from none to less than 7% under normal conditions. Note that the reduced sampling plan is the same as the single sampling plan being replaced.

Table 7.2 only offers QSSs giving improved protection for single sampling plans with zero acceptance numbers. However, the same approach can be used with any single or double sampling plan. Option 7 of the program QSS selects QSSs offering improved protection. Screen 7.13 shows the input required to select a QSS offering improved protection as a replacement for the single sampling plan n=50 and a=1. The QSS selected is n_r=50, a_r=1, n_t=81, a_t=1 and

$s_t=0$. Option 7 uses the specified sampling plan for the reduced inspection. It then selects the tightened inspection with the smallest possible LTPD satisfying the four restrictions on the maximum chances of making errors and the rates of convergence.

Screen 7.13: Input to Select Improved Protection QSS (QSS)

```
*****************************************************************************
***                                                                      ***
***        QSS Option 7 - Select QSS With Specified Reduced Inspection    ***
***                                                                      ***
*****************************************************************************

   This option constructs a QSS using the specified reduced inspection from
   among those satisfying the following conditions:

      Maximum chance of rejection at AQL is 0.15
      Rate of convergence at AQL is at least 50%
      Maximum chance of acceptance at LTPD is 0.3
      Rate of convergence at LTPD is at least 70%

   The standard definitions of AQL (95% chance of acceptance),
   LTPD (10% chance of acceptance) and the values given above are used
   unless indicated otherwise.

TALLY? (1=defectives,2=defects)          --> 1
REDUCED INSP.? ("S"=single, "D"=double) --> s
ENTER SAMPLE SIZE REDUCED INSP.   (nr) --> 50
ENTER ACCEPT NUMBER REDUCED INSP.  (ar) --> 1
USE DEFAULT DEFINITIONS? ("y" or "n")    --> y
```

Exercise

(7.14) Select a QSS offering improved protection at the same cost as the double sampling plan n1=32, a1=0, r1=2, n2=32 and a2=1. Assume defectives are tallied. How do the AQL and LTPD change? How is the sample size affected at 0.65% defective?

7.8   Monitoring Processes

Frequently one monitors processes by inspecting small numbers of samples on a routine basis. In Chapter 4 it was shown that if there is a possibility of a major process breakdown, even if only a small chance, such monitoring is the most economical choice. Process monitoring can catch major problems and some more minor problems. Also, it occasionally falsely signals a problem when none exists. One issue is what to do when a problem is detected. Obviously the problem should be fixed. However, one may also want to verify the existence of the problem first to avoid reacting to false signals. One should also confirm the effectiveness of the fix. In addition, the level of inspection should be increased until a fix can be implemented.

These issues can be addressed using QSSs. Suppose one is monitoring the process by selecting 13 samples from each lot. Any defective found is taken to signify a problem. Once a problem is detected, increased inspection should be implemented. Suppose it is decided that subsequent lots should be inspected using the single sampling plan n=200 and a=5. Further, inspection is to continue until the problem is confirmed to have been fixed. A sample of 200 units free of defects is required to confirm a fix. The above procedure is equivalent to using the QSS n_r=13, a_r=0, n_t=200, a_t=5 and s_t=0.

The above procedure can falsely signal a problem when none exists. As a result, it was decided to confirm a problem by selecting a second set of samples whenever a single defect was found. It was decided to use the same single sampling plan n=200 and a=5 to confirm a problem as used to inspect subsequent lots. The procedure now is equivalent to the QSS $n1_r$=13, $a1_r$=0, $r1_r$=2, $n2_r$=200, $a2_r$=6, n_t=200, a_t=5 and s_t=0. The performance of this monitoring procedure can be evaluated using the program QSS. The summary statistics of this QSS are:

AQL	=	1.42
LTPD	=	4.59
Maximum chance of rejection at AQL	=	0.0671
Rate of convergence at AQL	=	5.7%
Maximum chance of acceptance at LTPD	=	0.559
Rate of convergence at LTPD	=	44.1%

This example demonstrates that there are applications for QSSs that do not satisfying the four restrictions on the maximum chance of making errors and the rates of convergence. The purpose of these restrictions is to select QSSs that can be used interchangeably with single and double sampling plans. However, QSSs also have other uses.

7.9   Comparison to Chain Sampling Plans

Chain sampling is another type of sampling plan that, like QSSs, is only applicable to the inspection of a series of lots. Take as an example, the chain sampling plan n=20, a1=0, r1=2, i=3 and a2=1. This chain sampling plan selects n=20 samples from each lot. Lots whose samples contains a1=0 defectives are accepted. Lots whose sample contains r1=2 or more defectives are rejected. Otherwise the total number of defectives in the current lot and the i-1=2 previous lots is tallied. If this tally is less than or equal to a2=1, the lot is accepted. Otherwise it is rejected. This chain sampling plan is equivalent to the double sampling plan with n1=n and n2=(i-1)×n except that the second sample is selected from previous lots instead of the current lot.

The protection provided by chain sampling plans depends on the quality of previous lots as well as the current lot. Therefore stationary and transitive OC curves can be constructed. The stationary OC curve is the same as that of the double sampling plan with n1=n and n2=(i-1)×n.

Taylor (1992b and 1993) compares chain sampling plans with QSSs selected to have similar stationary and transitive OC curves. In all cases, QSSs provided equivalent protection for far fewer units inspected. Chain sampling does, however, offer one advantage. Procedurally, it requires one sample from each lot where the sample is of a constant size. One sample per lot is generally a requirement when units are tested as a group. A constant sample size may be required if the number of units tested per group is limited. QSSs using a single sampling plan for reduced inspection whose sample size is equal to the sample size of the tightened inspection also satisfies these requirements. Under these restrictions, neither chain sampling plans nor QSSs has any clear advantage. As a result of the superiority of QSSs, chain sampling plans are not covered in this book.

7.10   Summary

QSS are not always applicable. They can only be used to inspect a series of lots. When applicable, QSSs offer nearly the same protection as single and double sampling plans for far fewer units inspected. QSSs reduce the number of units inspected by 25% to 40%

compared to double sampling plans and by 50% to 75% compared to single sampling plans. QSSs have increased risks of accepting bad lots and rejecting good lots following changes in quality. Using the procedures given, QSSs can be selected where the increased chance of making errors is limited in size and only remains significant for the first lot following a change.

The protection provided by QSSs depends on the quality of previous lots along with the quality of the current lot. Both stationary OC curves (constant quality) and transitive OC curves (changing quality) are required to describe their protection. Only Type-B OC curves apply. These OC curves can be summarized using the AQL, LTPD, maximum chance of rejection at the AQL, rate of convergence at the AQL, maximum chance of acceptance at the LTPD and the rate of convergence at the LTPD. QSSs prove to be superior to chain sampling.

Procedures are given for selecting QSSs of a specified AQL and LTPD that limit the maximum chances of making errors and ensure rapid convergence. Such QSSs offer a safe alternative to single and double sampling plans. QSSs can also be used to provide improved protection at the same cost. This is the only alternative available to single sampling plans with accept numbers of zero. A third use of QSSs is for process monitoring. QSSs can ensure fixes are verified, allow for increased inspection until such verification, and verify the existence of a problem to screen out false signals.

References

Taylor, Wayne A. (1992a). "Quick Switching Systems: Part 1 - Evaluating." Submitted for publication.

Taylor, Wayne A. (1992b). "Quick Switching Systems: Part 2 - Selecting and Comparing." Submitted for publication.

Taylor, Wayne A. (1992c). "A Program for Selecting Quick Switching Systems." Submitted for publication.

Taylor, Wayne A. (1993). "Quick Switching Systems for Acceptance Sampling." Transactions of 1993 Annual Quality Congress, ASQC, Milwaukee, Wisconsin.

Mil-Std-105E

Mil-Std-105E is the most commonly used table of attributes sampling plans. It is frequently referenced in government contracts. It is also commonly referenced in procedures and standards published by many different regulatory agencies, such as the FDA, and by numerous standard committees.

Mil-Std-105E contains tables of single, double and multiple sampling plans. These sampling plans are indexed by AQL and level of inspection. The OC curves, percentiles, AOQLs and ASN curves are given for the different sampling plans. Mil-Std-105E actually has three tables for each type of sampling plan. These tables are referred to as the normal, reduced, and tightened inspection tables. Mil-Std-105E also provides a set of rules for switching between the normal, reduced, and tightened sampling plans. The result is a switching system consisting of three different sampling plans.

This chapter has three objectives:

- Learn to use the indexing system in Mil-Std-105E.

- Learn to use the switching system in Mil-Std-105E.

- Learn to replace Mil-Std-105E double and multiple sampling plans with more efficient double sampling plans and QSSs.

- Learn how to replace the Mil-Std-105E switching rules with simpler and more effective QSSs.

8.1   The Mil-Std-105E Indexing System

Mil-Std-105E sampling plans are indexed by AQLs and levels of inspection. For example, a specification might state: "Inspect per Mil-Std-105E using a L-II inspection with an AQL of 1.0%." L-II refers to general inspection level II. Determining the correct sampling plan to use also requires knowledge of the lot size. The procedure for determining the sampling plan to use is:

- Determine the sample size letter code from Table 1 of Mil-Std-105E. The sample size letter code depends on the level of inspection and lot size. This table is reproduced in Table 8.1 on page 174.

- Determine the single sampling plan to use from Table II-A of Mil-Std-105E. The sampling plan depends on the AQL and sample size letter code. This table is reproduced in Table 8.2 on page 175.

- If one lands on an arrow, follow the arrow up or down to the first pair of accept-reject numbers. Then look up the corresponding sample size from the sample size column.

Example 1: Assume a L-II inspection with an AQL of 1.0% is specified and that the lot sizes range from 4,000 to 8,000 units. Using Table 1 of Mil-Std-105E, the sample size letter code is L. From Table II-A of Mil-Std-105E, the single sampling plan is n=200 and a=5.

Example 2: Assume a S-3 inspection with an AQL of 1.0% is specified and that the lot sizes range from 4,000 to 8,000 units. Using Table 1 of Mil-Std-105E, the sample size letter code is G. This combination of sample size letter code and AQL results in a downward arrow. This arrow leads down one row to the single sampling plan n=50 and a=1.

Example 3: Assume a S-4 inspection with an AQL of 1.0% is specified and that the lot sizes range from 600 to 2,000 units. Using Table 1 of Mil-Std-105E, the sample size letter code is F for lots of 1,200 or fewer units and G for lots of 1,201 or more units. From Table II-A of Mil-Std-105E, letter F results in the single sampling plan n=13 and a=0. For letter G, the single sampling plan is n=50 and a=1. The single sampling plan n=13 and a=0 should be used for lots of 1,200 or fewer units and the single sampling plan n=50 and a=1 should be used for lots of 1,201 or more units.

Mil-Std-105E states the following about the level of inspection:

> "Three inspection levels: I, II, and III, are given in Table I for general use. Normally, Inspection Level II is used. However, Inspection Level I may be used when less discrimination is needed, or Level III may be used for greater discrimination. Four additional special levels: S-1, S-2, S-3, and S-4 are given in the same table and may be used where relatively small sample sizes are necessary and large sampling risks can, or must be tolerated."

The higher the level of inspection, the more severe the sampling plan is, i.e., the lower the LTPD. The levels of inspection in increasing order are: S-1, S-2, S-3, S-4, L-I, L-II and L-III.

Mil-Std-105E is frequently specified in government contracts. Such contracts generally specify a receiving inspection per Mil-Std-105E with a specified level of inspection and AQL. The receiver is, of course, the government. Such contracts generally do not require that the producer use Mil-Std-105E. If you believe you cannot take advantage of improved double sampling plans and QSSs because of your contract, read it carefully. Further, inspecting your outgoing product with the same sampling plan used for the receiving inspection does not guarantee all lots will pass the receiving inspection. Lots at the AQL have a 5% chance of being rejected the first time inspected. Those lots released still have a 5% chance of rejection by the receiving inspection. Instead of repeating the inspection to be performed as part of receiving inspection, the producer should implement the controls and inspection necessary to ensure good quality product.

Mil-Std-105E can be used to select single, double and multiple sampling plans based on their protection. Mil-Std-105E contains a wide variety of such sampling plans organized by the protection they provide. The percentiles, AOQLs and ASN curves of these sampling plans are all provided. Tables are given for when defectives are tallied and for when defects are tallied. Using these tables, one can select a sampling plan providing a specified AQL and LTPD. One can also select a sampling plan providing a specified AQL and AOQL. However, before using the tables in Mil-Std-105E, one needs to be aware of how the Mil-Std-105E indexing system works.

Table 8.1: Mil-Std-105E Table I - Sample Size Letter Code

Lot or batch size			Special inspection levels				General inspection levels		
			S-1	S-2	S-3	S-4	I	II	III
2	to	8	A	A	A	A	A	A	B
9	to	15	A	A	A	A	A	B	C
16	to	25	A	A	B	B	B	C	D
26	to	50	A	B	B	C	C	D	E
51	to	90	B	B	C	C	C	E	F
91	to	150	B	B	C	D	D	F	G
151	to	280	B	C	D	E	E	G	H
281	to	500	B	C	D	E	F	H	J
501	to	1200	C	C	E	F	G	J	K
1201	to	3200	C	D	G	G	H	K	L
3201	to	10000	C	D	G	G	J	L	M
10001	to	35000	C	D	H	H	K	M	N
35001	to	150000	D	E	J	J	L	N	P
150001	to	500000	D	E	J	J	M	P	Q
500001	and	over	D	E	K	K	N	Q	R

Table 8.2: Mil-Std-105E Table II-A - Single sampling plans for normal inspection (Master table)

Acceptable Quality Levels (normal inspection)

Sample size code letter	Sample size	0.010	0.015	0.025	0.040	0.065	0.10	0.15	0.25	0.40	0.65	1.0	1.5	2.5	4.0	6.5	10	15	25	40	65	100	150	250	400	650	1000
		Ac Re	Ac Re	Ac Re	Ac Re	Ac Re	Ac Re	Ac Re	Ac Re	Ac Re	Ac Re	Ac Re	Ac Re	Ac Re	Ac Re	Ac Re	Ac Re	Ac Re	Ac Re	Ac Re	Ac Re	Ac Re	Ac Re	Ac Re	Ac Re	Ac Re	Ac Re
A	2	↓	↓	↓	↓	↓	↓	↓	↓	↓	↓	↓	↓	↓	↓	0 1	↑	↓	1 2	2 3	3 4	5 6	7 8	10 11	14 15	21 22	30 31
B	3	↓	↓	↓	↓	↓	↓	↓	↓	↓	↓	↓	↓	↓	0 1	↑	↓	1 2	2 3	3 4	5 6	7 8	10 11	14 15	21 22	30 31	44 45
C	5	↓	↓	↓	↓	↓	↓	↓	↓	↓	↓	↓	↓	0 1	↑	↓	1 2	2 3	3 4	5 6	7 8	10 11	14 15	21 22	30 31	44 45	↑
D	8	↓	↓	↓	↓	↓	↓	↓	↓	↓	↓	↓	0 1	↑	↓	1 2	2 3	3 4	5 6	7 8	10 11	14 15	21 22	30 31	44 45	↑	↑
E	13	↓	↓	↓	↓	↓	↓	↓	↓	↓	↓	0 1	↑	↓	1 2	2 3	3 4	5 6	7 8	10 11	14 15	21 22	30 31	44 45	↑	↑	↑
F	20	↓	↓	↓	↓	↓	↓	↓	↓	↓	0 1	↑	↓	1 2	2 3	3 4	5 6	7 8	10 11	14 15	21 22	30 31	44 45	↑	↑	↑	↑
G	32	↓	↓	↓	↓	↓	↓	↓	↓	0 1	↑	↓	1 2	2 3	3 4	5 6	7 8	10 11	14 15	21 22	30 31	44 45	↑	↑	↑	↑	↑
H	50	↓	↓	↓	↓	↓	↓	↓	0 1	↑	↓	1 2	2 3	3 4	5 6	7 8	10 11	14 15	21 22	30 31	44 45	↑	↑	↑	↑	↑	↑
J	80	↓	↓	↓	↓	↓	↓	0 1	↑	↓	1 2	2 3	3 4	5 6	7 8	10 11	14 15	21 22	30 31	44 45	↑	↑	↑	↑	↑	↑	↑
K	125	↓	↓	↓	↓	↓	0 1	↑	↓	1 2	2 3	3 4	5 6	7 8	10 11	14 15	21 22	30 31	44 45	↑	↑	↑	↑	↑	↑	↑	↑
L	200	↓	↓	↓	↓	0 1	↑	↓	1 2	2 3	3 4	5 6	7 8	10 11	14 15	21 22	30 31	44 45	↑	↑	↑	↑	↑	↑	↑	↑	↑
M	315	↓	↓	↓	0 1	↑	↓	1 2	2 3	3 4	5 6	7 8	10 11	14 15	21 22	30 31	44 45	↑	↑	↑	↑	↑	↑	↑	↑	↑	↑
N	500	↓	↓	0 1	↑	↓	1 2	2 3	3 4	5 6	7 8	10 11	14 15	21 22	30 31	44 45	↑	↑	↑	↑	↑	↑	↑	↑	↑	↑	↑
P	800	↓	0 1	↑	↓	1 2	2 3	3 4	5 6	7 8	10 11	14 15	21 22	30 31	44 45	↑	↑	↑	↑	↑	↑	↑	↑	↑	↑	↑	↑
Q	1250	0 1	↑	↓	1 2	2 3	3 4	5 6	7 8	10 11	14 15	21 22	30 31	44 45	↑	↑	↑	↑	↑	↑	↑	↑	↑	↑	↑	↑	↑
R	2000	↑	↓	1 2	2 3	3 4	5 6	7 8	10 11	14 15	21 22	30 31	44 45	↑	↑	↑	↑	↑	↑	↑	↑	↑	↑	↑	↑	↑	↑

↓ = Use first sampling plan below arrow. If sample size equals, or exceeds, batch size, do 100 percent inspection.

↑ = Use first sampling plan above arrow.

Ac = Acceptance number.

Re = Rejection number.

Table 8.3 gives the actual AQLs and LTPDs of the single sampling plans in Mil-Std-105E indexed under an AQL of 1.0%. Mil-Std-105E does not hold the probability of acceptance at the AQL constant. Instead, the probability of acceptance at the AQL steadily increases from 0.877 to 0.991. This results in the actual AQLs starting at 0.4%, increasing to 1.3% and then decreasing again. The actual AQL may vary from 60% below to 30% above the indexing AQL. In contrast, the single sampling plans in Table 3.3 have actual AQLs that closely approximate their indexed value.

**Table 8.3: Probability of Acceptance at AQL for
AQL = 1.0% Mil-Std-105E Sampling Plans**

Sample Size Letter Code	Sampling Plan	OC(1%)	Actual AQL	LTPD
E	n=13 a=0	0.877	0.394%	16.2%
H	n=50 a=1	0.911	0.715%	7.56%
J	n=80 a=2	0.953	1.03%	6.52%
K	n=125 a=3	0.963	1.10%	5.27%
L	n=200 a=5	0.984	1.31%	4.59%
M	n=315 a=7	0.985	1.27%	3.71%
N	n=500 a=10	0.987	1.24%	3.06%
P	n=800 a=14	0.983	1.16%	2.51%
Q	n=1250 a=21	0.991	1.19%	2.25%

The fact that the probability of acceptance varies so widely at the indexing AQL is a frequent source of problems. Suppose a process averages 0.8% defective. As part of selecting a sampling plan for this process, one might specify an AQL of 1% to ensure a low scrap rate. Using Mil-Std-105E and specifying an indexing AQL of 1.0% might result in the selection of the single sampling plan n=13 and a=0. This sampling plan would reject 10% of current production. This is certain to raise a storm of protest. What went wrong? The problem is that, while the single sampling plan n=13 and a=0 is indexed under an AQL of 1%, its actual AQL is 0.4%. This problem is easily avoided by examining   the   percentiles   and   OC   curves   given   in Mil-Std-105E.

Table 3.3 requires that the actual AQL closely approximate the indexing AQL. As a result, the indexing AQLs of many of the single sampling plans in Table 3.3 are different than their indexing AQLs in Mil-Std-105E. The single sampling plan n=13 and a=0 is indexed under an AQL of 1% in Mil-Std-105E. It has an actual AQL of 0.394%. In Table 3.3 it is indexed under an AQL of 0.4% instead.

Not all Mil-Std-105E single sampling plans can be found in Table 3.3. One such single sampling plan is n=200 and a=5. This sampling plan has an actual AQL of 1.31% which is not close to any of the indexed values. Instead, Table 3.3 contains the single sampling plan n=200 and a=4. It has an actual AQL of 0.990% and is indexed under an AQL of 1%. Table 3.3 also contains the single sampling plan n=200 and a=6. It has an actual AQL of 1.65% and is indexed under an AQL of 1.5%. Mil-Std-105E uses the following sequence of accept numbers: 0, 1, 2, 3, 5, 7, 10, 14 and 21. In contrast, Table 3.3 uses the sequence 0, 1, 2, 3, 4, 6, 8, 12 and 18.

Why does Mil-Std-105E change sampling plans as lot size increases? To understand why, one must examine the effect of changing the lot size. Increasing the lot size causes the sample size letter code to increase. As shown in Table 8.3, this results in a steeper OC curve providing increased protection. The probability of acceptance increases at the AQL while the LTPD decreases. Mil-Std-105E was purposely designed to provide better protection for larger lots than for smaller lots. It makes perfect sense to invest more in the inspection of large lots since more is at risk. This is especially true when the lots are all the same product. However, there are many other factors effecting the selection of sampling plans besides lot size. Generally, one would want better protection for a small lot of cars than for a large lot of nails.

The fact that Mil-Std-105E changes sampling plans as lot size increases is a source of much confusion. Many believe that Mil-Std-105E changes sampling plans because more samples are required to obtain equivalent protection for larger lots. Nothing is farther from the truth. In Chapter 2 we saw that a sample of 50 units provides the same protection for a lot of 500 units as for a lot of 500,000 units. If it were desired to keep the protection constant, the sampling plan need not change as lot size increases.

Do not forget that Mil-Std-105E was written for use in receiving inspections performed by the government. In this light, there is justification for the practice of concentrating inspection effort on larger lots. However, the tables in Mil-Std-105E are also commonly used to select sampling plans based on the protection they provide. The indexing system in Mil-Std-105E complicates its use in this manner. Tables 3.3, 6.3, and 7.2 are much better suited for this purpose.

While there is some justification for changing sampling plans as the lot size increases, it is impossible to justify the large changes in sample sizes that sometimes result from small changes in the lot size. For a S-3 inspection with an AQL of 1.0%, a change in lot size from 1200 to 1201 units results in the sample size increasing from 13 to 50. When lot size varies, use the average lot size to determine the sampling plan and then use this sampling plan for all lots.

The sampling plans in Table 8.2 are for what Mil-Std-105E refers to as normal inspections. Mil-Std-105E also gives tables of tightened and reduced sampling plans along with a set of rules for switching between them. These switching systems can lower inspection costs. These switching rules are the topic of Section 8.3.

Exercises

(8.1)   Determine which sampling plan to use if a L-I inspection is specified with an AQL of 0.40% where lot sizes range from 5,000 to 10,000 units.

(8.2)   Determine the probability of acceptance for this sampling plan at 0.40% along with its AQL and LTPD. How is this sampling plan indexed in Table 3.3 (page 87)?

8.2   Matching Double Sampling Plans and QSSs

Mil-Std-105E provides matching double and multiple sampling plans for each of its single sampling plans. Use of these matching plans can significantly reduce the amount of inspection. However, the double sampling plans in Mil-Std-105E were selected under the restriction $n1=n2=n/1.6$ where n is the sample size of the single sampling plan being matched. Similar restrictions were placed on the sample sizes of multiple sampling plans. These restrictions reduce the savings in inspection costs.

Table 8.4 (page 181) provides an alternative set of matching double sampling plans. They can reduce the amount of inspection below that of the double sampling plans given in Mil-Std-105E. In most cases, these double sampling plans reduce the amount of inspection below that of the multiple sampling plans in Mil-Std-105E. This table was generated using the program DOUBLE. The ASN was minimized at the $^{AQL}/2$.

Similarly, Table 8.5 (page 183) provides matching QSSs. The QSSs in Table 8.5 satisfy the restrictions:

Maximum chance of rejection at AQL	$\leq$ 0.15
Rate of convergence at AQL	$\geq$ 50%
Maximum chance of acceptance at LTPD	$\leq$ 0.3
Rate of convergence at LTPD	$\geq$ 70%

These restrictions allow a tripling of the risk of making an error for the first lot following a change in quality with most of the increased risk disappearing by the second lot. While not identical in the protection provided, little is compromised using these QSSs. In return, they can result in sizable reductions in inspection costs. This table was generated using the program QSS. The ASN was minimized at the $^{AQL}/2$.

Example 1: Assume the single sampling plan n=200 and a=5 is specified. The matching double sampling plan per Table 8.4 is n1=90, a2=1, r1=6, n2=180 and a2=6. The matching QSS per Table 8.5 is $n1_r$=32, $a1_r$=0, $r1_r$=4, $n2_r$=125, $a2_r$=4, n_t=160, a_t=3 and s_t=2. Figure 8.1 shows the ASN curves of these two sampling plans along with the ASN curves of the matching double and multiple sampling plans given in Mil-Std-105E. The double sampling plan given in Mil-Std-105E is n1=125, a1=2, r1=5, n2=125 and a2=6.

When comparing ASN curves, one should heavily weight the sections of these curves ranging from 0% defective to the AQL. Most lots inspected generally fall in this region. These sections have been boxed to highlight them. The QSS requires the least amount of inspection. It can reduce the amount of inspection by 60% to 84% when compared to the original single sampling plan. The double sampling plan from Table 8.4 and the multiple sampling plan from Mil-Std-105E requires about the same amount of inspection. Both are superior to the double sampling plan from Mil-Std-105E. Example 1 shows what happens when the original single sampling plan has an accept number of 2 or greater.

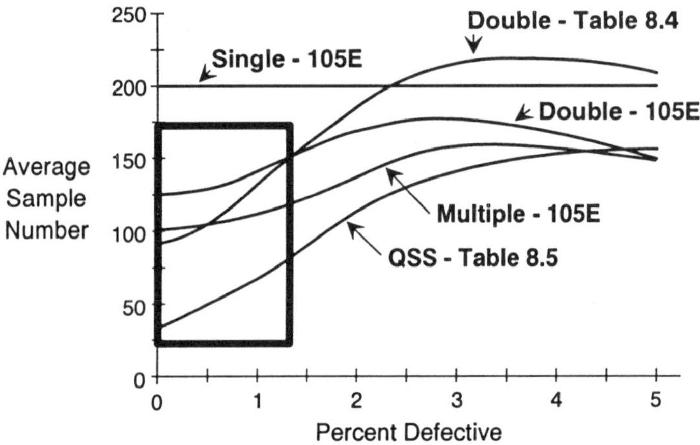

**Figure 8.1: ASN Curves of Matching Plans for
Single Sampling Plan n=200 and a=5.**

Example 2: For the sampling plan n=50 and a=1, the matching double sampling plan in Table 8.4 is n1=32, a2=0, r1=2, n2=32 and a2=1. This same double sampling plan is given in Mil-Std-105E. The matching QSS per Table 8.5 is $n1_r$=20, $a1_r$=0, $r1_r$=2, $n2_r$=40, $a2_r$=1, n_t=50, a_t=1 and s_t=0. This QSS offers considerable reduction in the average sample number compared to the matching double and multiple sampling plans. Example 2 is representative of finding replacements for single sampling plans with accept numbers of 1.

Example 3: For the single sampling plan n=13 and a=0, neither Mil-Std-105E nor Table 8.4 provides a matching sampling plan. Table 8.5 gives the QSS n_r=13, a_r=0, n_t=26, a_t=0 and s_t=0. Instead of offering the same protection at a reduced cost, this QSS offers improved protection at the same cost. It improves the LTPD from 16.2% to 9.27%. Example 3 is representative of matching single sampling plans with zero accept numbers.

Exercises

(8.3)   Find the matching double sampling plan and QSS for the single sampling plan n=80 and a=2. Compare the AQL, LTPD, ASN(0%) and ASN(1%) for these sampling plans.

(8.4)   Find the matching double sampling plan and QSS for the single sampling plan n=20 and a=0. Compare the AQL, LTPD, ASN(0%) and ASN(0.25%) of these sampling plans.

Table 8.4: Matching Double Sampling Plans for Mil-Std-105E Normal Inspections

Sample Size	Acceptance Number										
	0	1	2	3	5	7	10	14	21	30	44
3	*	2/(0,2) 2/(1,2)	2/(0,3) 2/(3,4)	1/(0,3) 5/(5,6)	1/(0,4) 3/(6,7)	1/(1,5) 5/(11,12)	1/(1,6) 3/(13,14)	1/(3,10) 5/(23,24)	1/(5,10) 5/(39,40)	1/(7,15) 3/(39,40)	1/(12,19) 4/(67,68)
5	*	3/(0,2) 3/(1,2)	3/(0,3) 3/(3,4)	2/(0,3) 5/(4,5)	2/(1,4) 10/(10,11)	2/(1,6) 4/(8,9)	2/(2,7) 5/(13,14)	2/(4,8) 6/(21,22)	2/(6,14) 4/(24,25)	1/(3,12) 5/(35,36)	1/(6,14) 6/(60,61)
8	*	5/(0,2) 5/(1,2)	4/(0,3) 8/(3,4)	3/(0,4) 9/(4,5)	4/(1,5) 6/(6,7)	3/(1,6) 7/(8,9)	3/(2,7) 8/(13,14)	3/(3,9) 7/(17,18)	2/(3,11) 8/(25,26)	2/(5,12) 11/(46,47)	2/(8,18) 8/(53,54)
13	*	8/(0,2) 8/(1,2)	6/(0,3) 14/(3,4)	5/(0,4) 13/(4,5)	6/(1,5) 11/(6,7)	5/(1,6) 11/(8,9)	5/(2,8) 10/(11,12)	4/(2,8) 10/(15,16)	4/(4,11) 12/(25,26)	4/(7,16) 15/(41,42)	4/(11,20) 16/(64,65)
20	*	13/(0,2) 13/(1,2)	9/(0,3) 23/(3,4)	8/(0,3) 20/(4,5)	9/(1,6) 18/(6,7)	7/(1,7) 23/(9,10)	7/(2,8) 23/(13,14)	7/(3,10) 20/(18,19)	5/(3,11) 20/(25,26)	-	-
32	*	20/(0,2) 20/(1,2)	15/(0,3) 36/(3,4)	12/(0,3) 32/(4,5)	14/(1,6) 28/(6,7)	11/(1,7) 36/(9,10)	11/(2,8) 36/(13,14)	11/(3,10) 32/(18,19)	8/(3,11) 32/(25,26)	-	-
50	*	32/(0,2) 32/(1,2)	24/(0,3) 56/(3,4)	19/(0,3) 50/(4,5)	23/(1,6) 46/(6,7)	18/(1,7) 56/(9,10)	18/(2,8) 56/(13,14)	18/(3,10) 50/(18,19)	13/(3,11) 50/(25,26)	-	-
80	*	50/(0,2) 50/(1,2)	38/(0,3) 90/(3,4)	30/(0,3) 80/(4,5)	36/(1,6) 72/(6,7)	28/(1,7) 90/(9,10)	28/(2,8) 90/(13,14)	28/(3,10) 80/(18,19)	20/(3,11) 80/(25,26)	-	-

* There is no matching double sampling plan.

Table 8.4: (continued)

Sample Size	Acceptance Number										
	0	1	2	3	5	7	10	14	21	30	44
125	*	80/(0,2) 80/(1,2)	60/(0,3) 140/(3,4)	48/(0,3) 125/(4,5)	55/(1,6) 110/(6,7)	45/(1,7) 140/(9,10)	45/(2,8) 140/(13,14)	45/(3,10) 125/(18,19)	32/(3,11) 125/(25,26)	-	-
200	*	125/(0,2) 125/(1,2)	94/(0,3) 225/(3,4)	75/(0,3) 200/(4,5)	90/(1,6) 180/(6,7)	70/(1,7) 225/(9,10)	70/(2,8) 225/(13,14)	70/(3,10) 200/(18,19)	50/(3,11) 200/(25,26)	-	-
315	*	200/(0,2) 200/(1,2)	150/(0,3) 350/(3,4)	120/(0,3) 315/(4,5)	140/(1,6) 280/(6,7)	110/(1,7) 350/(9,10)	110/(2,8) 360/(13,14)	110/(3,10) 320/(18,19)	80/(3,11) 315/(25,26)	-	-
500	*	315/(0,2) 315/(1,2)	240/(0,3) 550/(3,4)	190/(0,3) 500/(4,5)	225/(1,6) 450/(6,7)	175/(1,7) 550/(9,10)	180/(2,8) 550/(13,14)	175/(3,10) 500/(18,19)	125/(3,11) 500/(25,26)	-	-
800	*	500/(0,2) 500/(1,2)	375/(0,3) 900/(3,4)	300/(0,3) 800/(4,5)	360/(1,6) 700/(6,7)	280/(1,7) 900/(9,10)	280/(2,8) 900/(13,14)	280/(3,10) 800/(18,19)	200/(3,11) 800/(25,26)	-	-
1250	*	800/(0,2) 800/(1,2)	600/(0,3) 1425/(3,4)	470/(0,3) 1250/(4,5)	560/(1,6) 1150/(6,7)	440/(1,7) 1425/(9,10)	440/(2,8) 1425/(13,14)	440/(3,10) 1250/(18,19)	315/(3,11) 1250/(25,26)	-	-
2000	-	1250/(0,2) 1250/(1,2)	940/(0,3) 2250/(3,4)	750/(0,3) 2000/(4,5)	900/(1,6) 1800/(6,7)	700/(1,7) 2250(9,10)	700/(2,8) 2250(13,14)	700/(3,10) 2000/(18,19)	500/(3,11) 2000/(25,26)	-	-

* There is no matching double sampling plan.

Table 8.5: Matching Quick Switching Systems for Mil-Std-105E Normal Inspections

Sample Size	Acceptance Number										
	0	1	2	3	5	7	10	14	21	30	44
3	3/(0,1) * 6/(0,1) 0	1/(0,2) 2/(1,2) 3/(1,2) 0	1/(0,2) 2/(2,3) 3/(2,3) 1	1/(0,3) 2/(3,4) 3/(3,4) 1	1/(1,4) 2/(5,6) 3/(4,5) 2	1/(2,4) 2/(8,9) 2/(4,5) 2	1/(3,6) 2/(10,11) 3/(9,10) 6	1/(4,8) 1/(10,11) 3/(14,15) 9	1/(7,12) 2/(20,21) 3/(19,20) 15	1/(10,14) 2/(30,31) 3/(27,28) 22	1/(15,19) 3/(58,59) 3/(41,42) 34
5	5/(0,1) * 10/(0,1) 0	2/(0,2) 4/(1,2) 5/(1,2) 0	2/(0,3) 2/(2,3) 4/(1,2) 0	1/(0,3) 4/(3,4) 4/(2,3) 1	1/(0,3) 3/(5,6) 5/(4,5) 2	1/(0,4) 3/(6,7) 4/(5,6) 3	1/(1,4) 4/(12,13) 4/(7,8) 5	1/(2,6) 4/(14,15) 4/(10,11) 7	1/(3,8) 3/(18,19) 4/(15,16) 12	1/(5,10) 3/(25,26) 4/(22,23) 17	1/(8,14) 4/(44,45) 4/(33,34) 27
8	8/(0,1) * 16/(0,1) 0	3/(0,2) 6/(1,2) 8/(1,2) 0	2/(0,2) 6/(2,3) 6/(1,2) 0	2/(0,3) 3/(2,3) 7/(2,3) 1	2/(0,3) 4/(5,6) 6/(3,4) 2	2/(1,4) 4/(6,7) 7/(5,6) 3	2/(2,6) 6/(10,11) 7/(8,9) 5	1/(0,6) 7/(13,14) 6/(9,10) 6	1/(1,7) 4/(14,15) 5/(12,13) 9	1/(3,9) 12/(45,46) 6/(21,22) 17	1/(4,11) 5/(34,35) 6/(31,32) 26
13	13/(0,1) * 26/(0,1) 0	5/(0,2) 10/(1,2) 13/(1,2) 0	3/(0,2) 10/(2,3) 10/(1,2) 0	3/(0,3) 6/(2,3) 11/(2,3) 1	2/(0,4) 8/(4,5) 10/(3,4) 2	2/(0,4) 8/(6,7) 10/(5,6) 4	2/(1,5) 13/(10,11) 10/(7,8) 5	2/(1,6) 8/(12,13) 10/(10,11) 7	2/(2,9) 9/(18,19) 10/(15,16) 11	2/(4,9) 13/(33,34) 8/(17,18) 13	1/(2,8) 11/(41,42) 10/(32,33) 26
20	20/(0,1) * 40/(0,1) 0	8/(0,2) 16/(1,2) 20/(1,2) 0	5/(0,2) 16/(2,3) 16/(1,2) 0	4/(0,3) 10/(2,3) 16/(2,3) 1	3/(0,4) 13/(4,5) 16/(3,4) 2	3/(0,4) 13/(6,7) 16/(5,6) 4	3/(1,5) 20/(10,11) 16/(7,8) 5	3/(1,6) 13/(12,13) 16/(10,11) 7	3/(2,7) 13/(18,19) 16/(15,16) 12	-	-

* Provide improved protection.

Table 8.5: (continued)

Sample Size	Acceptance Number										
	0	1	2	3	5	7	10	14	21	30	44
32	32/(0,1) * 64/(0,1) 0	13/(0,2) 25/(1,2) 32/(1,2) 0	8/(0,2) 25/(2,3) 25/(1,2) 0	6/(0,3) 16/(2,3) 25/(2,3) 1	5/(0,4) 20/(4,5) 25/(3,4) 2	5/(0,4) 20/(6,7) 25/(5,6) 4	5/(1,5) 32/(10,11) 25/(7,8) 5	5/(1,6) 20/(12,13) 25/(10,11) 7	5/(2,7) 20/(18,19) 25/(15,16) 12	-	-
50	50/(0,1) * 100/(0,1) 0	20/(0,2) 40/(1,2) 50/(1,2) 0	13/(0,2) 40/(2,3) 40/(1,2) 0	10/(0,3) 25/(2,3) 40/(2,3) 1	8/(0,4) 32/(4,5) 40/(3,4) 2	8/(0,4) 32/(6,7) 40/(5,6) 4	8/(1,5) 50/(10,11) 40/(7,8) 5	8/(1,6) 32/(12,13) 40/(10,11) 7	8/(2,7) 32/(18,19) 40/(15,16) 12	-	-
80	80/(0,1) * 160/(0,1) 0	32/(0,2) 60/(1,2) 80/(1,2) 0	20/(0,2) 60/(2,3) 60/(1,2) 0	16/(0,3) 40/(2,3) 60/(2,3) 1	13/(0,4) 50/(4,5) 60/(3,4) 2	13/(0,4) 50/(6,7) 65/(5,6) 4	13/(1,5) 80/(10,11) 60/(7,8) 5	13/(1,6) 50/(12,13) 60/(10,11) 7	13/(2,7) 50/(18,19) 60/(15,16) 12	-	-
125	125/(0,1) * 250/(0,1) 0	50/(0,2) 100/(1,2) 125/(1,2) 0	32/(0,2) 100/(2,3) 100/(1,2) 0	25/(0,3) 60/(2,3) 100/(2,3) 1	20/(0,4) 80/(4,5) 100/(3,4) 2	20/(0,4) 80/(6,7) 100/(5,6) 4	20/(1,5) 125/(10,11) 100/(7,8) 5	20/(1,6) 80/(12,13) 100/(10,11) 7	20/(2,7) 80/(18,19) 100/(15,16) 12	-	-
200	200/(0,1) * 400/(0,1) 0	80/(0,2) 160/(1,2) 200/(1,2) 0	50/(0,2) 160/(2,3) 160/(1,2) 0	40/(0,3) 100/(2,3) 160/(2,3) 1	32/(0,4) 125/(4,5) 160/(3,4) 2	32/(0,4) 125/(6,7) 160/(5,6) 4	32/(1,5) 200/(10,11) 160/(7,8) 5	32/(1,6) 125/(12,13) 160/(10,11) 7	32/(2,7) 125/(18,19) 160/(15,16) 12	-	-

* Provide improved protection.

Table 8.5: (continued)

Sample Size	Acceptance Number										
	0	1	2	3	5	7	10	14	21	30	44
315	315/(0,1) * 630/(0,1) 0	125/(0,2) 250/(1,2) 315/(1,2) 0	80/(0,2) 250/(2,3) 250/(1,2) 0	60/(0,3) 160/(2,3) 250/(2,3) 1	50/(0,4) 200/(4,5) 250/(3,4) 2	50/(0,4) 200/(6,7) 250/(5,6) 4	50/(1,5) 315/(10,11) 250/(7,8) 5	50/(1,6) 200/(12,13) 250/(10,11) 7	50/(2,7) 200/(18,19) 250/(15,16) 12	-	-
500	500/(0,1) * 1000/(0,1) 0	200/(0,2) 400/(1,2) 500/(1,2) 0	125/(0,2) 400/(2,3) 400/(1,2) 0	100/(0,3) 250/(2,3) 400/(2,3) 1	80/(0,4) 315/(4,5) 400/(3,4) 2	80/(0,4) 315/(6,7) 400/(5,6) 4	80/(1,5) 500/(10,11) 400/(7,8) 5	80/(1,6) 315/(12,13) 400/(10,11) 7	80/(2,7) 315/(18,19) 400/(15,16) 12	-	-
800	800/(0,1) * 1600/(0,1) 0	315/(0,2) 625/(1,2) 800/(1,2) 0	200/(0,2) 625/(2,3) 625/(1,2) 0	160/(0,3) 400/(2,3) 625/(2,3) 1	125/(0,4) 500/(4,5) 625/(3,4) 2	125/(0,4) 500/(6,7) 625/(5,6) 4	125/(1,5) 800/(10,11) 625/(7,8) 5	125/(1,6) 500/(12,13) 625/(10,11) 7	125/(2,7) 500/(18,19) 625/(15,16) 12	-	-
1250	1250/(0,1) * 2500/(0,1) 0	500/(0,2) 1000/(1,2) 1250/(1,2) 0	315/(0,2) 1000/(2,3) 1000/(1,2) 0	250/(0,3) 625/(2,3) 1000/(2,3) 1	200/(0,4) 800/(4,5) 1000(3,4) 2	200/(0,4) 800/(6,7) 1000/(5,6) 4	200/(1,5) 1250/(10,11) 1000/(7,8) 5	200/(1,6) 800/(12,13) 1000/(10,11) 7	200/(2,7) 800/(18,19) 1000/(15,16) 12	-	-
2000	2000/(0,1) * 4000/(0,1) 0	800/(0,2) 1600/(1,2) 2000/(1,2) 0	500/(0,2) 1600/(2,3) 1600/(1,2) 0	400/(0,3) 1000/(2,3) 1600/(2,3) 1	315/(0,4) 1250/(4,5) 1600/(3,4) 2	315/(0,4) 1250/(6,7) 1600/(5,6) 4	315/(1,5) 2000/(10,11) 1600/(7,8) 5	315/(1,6) 1250/(12,13) 1600/(10,11) 7	315/(2,7) 1250/(18,19) 1600/(15,16) 12	-	-

* Provide improved protection.

8.3   Mil-Std-105E Switching Rules

Mil-Std-105E contains normal, reduced, and tightened sampling plans. It also provides a set of rules for switching between these plans. The switching rules are shown in Figure 8.2. Start using the normal inspection. The switching rules in Mil-Std-105E primarily depend on the number of recent lots accepted or rejected. For example, the rule for switching from normal to tightened is to switch if 2 of the last 5 lots are rejected while in normal inspection.

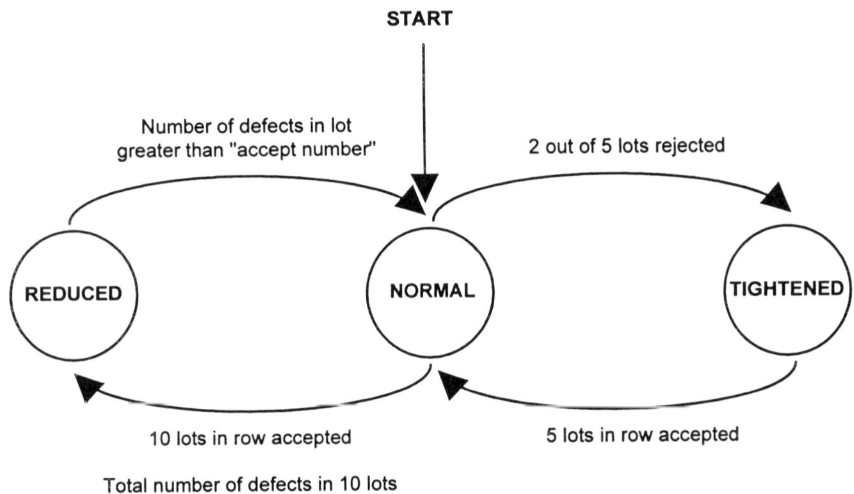

Figure 8.2: Mil-Std-105E Switching Rules

Switching from normal to reduced requires the acceptance of ten consecutive lots. It also requires the total number of defects in these ten lots to be less than or equal to the appropriate limit number. These limit numbers are given in Table VIII of Mil-Std-105E. The reduced inspections have reject numbers that are greater than the accept number plus one. For example, one of the reduced inspection single sampling plans is n=80, a=2 and r=5. Lots with a=2 or fewer defects are accepted and one is to remain on reduced inspection. Lots with 3 or 4 defects are accepted but one must switch to normal inspection. Lots with r=5 or more defects are rejected and one must switch to normal inspection. Mil-Std-105E also contains provisions for discontinuing production if 10 consecutive lots remain on tightened inspection.

The switching rules in Mil-Std-105E are not considered optional. ANSI/ASQC Standard Z1.4-1981, the equivalent civilian standard, states that:

> "Occasionally specific individual plans are selected from the standard and used without the switching rules. This is not the intended application of the ANSI Z1.4 system and its use in this way should not be referred to as inspection under ANSI Z1.4."

Nevertheless, in most applications in private industry, the switching rules are not used. This practice is encouraged by the fact that Mil-Std-105E only includes information on the protection provided by the individual sampling plans. ANSI Z1.4 corrects this deficiency by including the stationary OC curves when using the Mil-Std-105E switching rules.

Example 1: Assume a L-II inspection with an AQL of 1.0% is specified and that the lot sizes range from 4,000 to 8,000 units. The sample size letter code is L. The normal inspection is n=200 and a=5. The corresponding tightened inspection is n=200 and a=3. The reduced inspection is n=80, a=2 and r=5. The means to use the single sampling plan n=80 and a=4 to make accept/reject decisions and to switch to normal if 3 or more defects are found in a lot. The limit number for switching from normal to tightened is 14. This switching system has an AQL of 1.23% and a LTPD of 3.34%. The stationary OC curve is shown in Figure 8.3 along with the OC curves of the individual sampling plans.

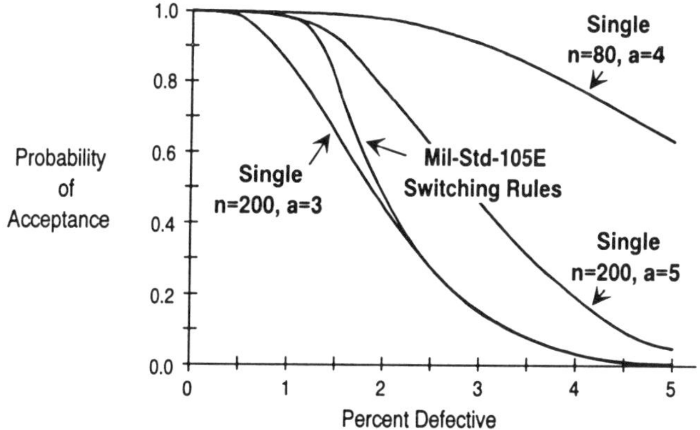

Figure 8.3: Stationary OC Curve of Switching Rules Along With OC Curves of Individual Sampling Plans for Sample Size Letter Code = L, AQL = 1.0%

Why should one use the Mil-Std-105E switching rules? The stock answer is that they provide improved protection at a reduced cost. The improved protection is provided by the tightened sampling plan. The reduced cost is provided by the reduced sampling plan. Certainly, the reduced cost is true. When the process average is at or below the AQL, most inspection is performed using the reduced inspection. This reduces the sample size from 200 to 80.

The part about improved protection in only true under stationary conditions. The LTPD of the normal sampling plan n=200 and a=5 is 4.64%. Use of the switching rules improves this to 3.34%. However, following a jump from good quality to quality near the LTPD, the first lot has a 68.6% chance of acceptance. Further, there is a 27.7% chance of failing to switch to normal. If it does switch to normal, 2 more lots must be rejected before switching to the improved protection of the tightened sampling plan. Use of the switching rules results in a significant degradation of the protection following such a change in quality. The protection does not improve beyond that offered by the normal sampling plan until at least the fourth lot following the change. The situation is much worse following a jump from poor quality as a minimum of 15 lots is required to switch to reduced inspection starting from tightened.

The truth is that use of the Mil-Std-105E switching rules can result in a significant degradation of protection during periods of changing quality. Further, these switching rules are complex and therefore difficult to administrate. One is justified in avoiding their use. The desired reduction in sample size is better achieved using matching double sampling plans and QSSs.

If one uses the Mil-Std-105E switching rules, one should consider using the procedure given in Taylor (1992) to obtain an equivalent QSS. This procedure consists of forming a QSS using the Mil-Std-105E reduced and tightened inspections and then selecting switch numbers to provide equivalent protection. Applying this procedure to example 1 results in the following QSS: n_r=80, a_r=4, s_r=4, n_t=200, a_t=3 and s_t=0. This QSS has one additional parameter s_r. For this QSS, the rule for switching from reduced to tightened changes to: switch if s_t or more defects are found in the lot. This QSS has an AQL of 1.26% and a LTPD of 3.33% compared to values of 1.23% and 3.34% for the Mil-Std-105E switching rules. The stationary OC curve of this QSS is nearly identical to that obtained using the Mil-Std-105E switching rules. While this QSS still has greatly increased chances of making errors following a jump in quality, it converges much more rapidly and is simpler to use.

8.4   Summary

Mil-Std-105E is the most commonly used table of attributes sampling plans. It contains tables of single, double and multiple sampling plans. It also contains tables and plots describing the protection provided by these sampling plans. Mil-Std-105E further contains matching reduced, normal, and tightened sampling plans along with a set of rules for switching between the sampling plans. Mil-Std-105E requires the use of these switching rules. Despite this, Mil-Std-105E is commonly used in private industry for selecting single and double sampling plans for stand alone use.

Mil-Std-105E was written for receiving inspections performed by the military. One must specify the level of inspection and AQL. The resulting sampling plan also depends on the lot size. Mil-Std-105E increases the protection as lot size increases in order to concentrate inspection on larger lots. As a result, the probability of acceptance is not held constant at the indexing AQLs. At the specified AQL, the probability of acceptance can range from 0.877 up to 0.991. While there is some economic justification for this practice, it complicates the use of the Mil-Std-105E for selecting single and double sampling plans for stand alone use. One should be sure to carefully examine the OC curve of the selected plan and not just rely on the specified AQL.

Tables 8.4 and 8.5 provide matching double sampling plans and QSSs for the Mil-Std-105E sampling plans. These sampling plans provide equivalent protection while reducing inspection costs. The double sampling plans are more efficient than those given in Mil-Std-105E because they do not place any restrictions on the parameters. They frequently are as efficient as the multiple sampling plans per Mil-Std-105E.

Care should be taken when using the Mil-Std-105E switching rules. Their use can result in significant degradation of protection during periods of changing quality. A better way of reducing sample size is to use the matching double sampling plans and QSSs. An alternative is to use a QSS consisting of the reduced and tightened sampling plans with switch numbers selected to provide equivalent protection. While this QSS will still suffer from degraded protection during periods of changing quality, it converges much more rapidly and is simpler to use.

References

ANSI/ASQC Z1.4. (1981). "Sampling Procedures and Tables for Inspection By Attributes." American Society for Quality Control, Milwaukee, WI.

Mil-Std-105E. (1989). "Sampling Procedures and Tables for Inspection By Attributes." U.S. Government Printing Office, Washington, D.C.

Taylor, Wayne A. (1992), "Quick Switching Systems: Part 2 - Selection and Comparison." Submitted for publication.

Chapter

9

Variables
Sampling Plans

When the characteristic being inspected can be measured, special sampling plans called variables sampling plans can be used. Variables sampling plans can dramatically reduce the number of units inspected. However, before variables sampling plans can be applied, certain assumptions must be verified.

This chapter has three objectives:

- Learn to make accept and reject decisions using variables sampling plans.

- Learn to evaluate the protection provided by variables sampling plans.

- Learn to select variables sampling plans based on the protection they provide.

9.1   The Normal Distribution

Variables sampling plans require measurements such as fill volumes, seal strengths and weights. Such data is called variables data. Variables sampling plans use such data to estimate the percent defective of a lot and to make an accept or reject decision. This requires knowing the manner in which these units are distributed. The variables sampling plans in this book will assume that the measurements follow the normal distribution.

Histograms are used to view the distribution of measurements. Figure 9.1 shows a histogram of 100 fill volume measurements.

Histograms are frequently shaped like symmetrical bell-shaped curves called normal curves. Figure 9.1 shows the normal curve superimposed over the histogram. This normal curve can be thought of as a histogram that has been smoothed out.

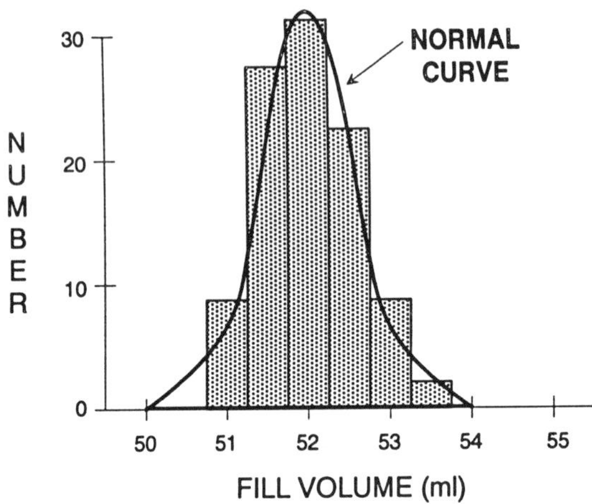

Figure 9.1: Histogram of 100 Fill Volumes

While all normal curves have the same general shape, their centers and widths vary. Figure 9.2 shows several normal curves. The center of the normal curve is called the mean and denoted μ. The width is described using the standard deviation σ. The width of the normal curve is approximately six times the standard deviation.

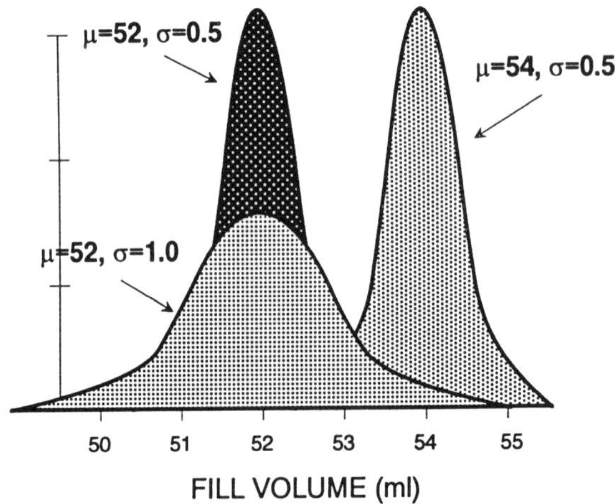

Figure 9.2: Different Normal Curves

If μ and σ are known, the percent defective can be estimated. Figure 9.3 shows a normal curve with μ=53.8 and σ=0.59. The mean is two standard deviations from the upper specification limit. This results in 2.3% of the product falling above the upper specification limit.

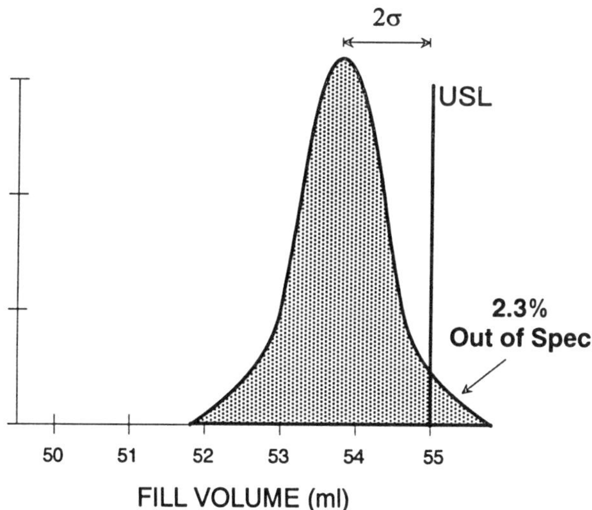

Figure 9.3: Estimating Percent Defective

To estimate the percent above an upper specification limit (USL), first calculate:

$$\text{Distance below upper spec. limit} \ = \ \frac{\text{USL} - \mu}{\sigma}$$

Then look up the corresponding percent defective in Table 9.1. Similarly, one can estimate the percent below a lower specification limit (LSL) by calculating the quantity below and looking up the corresponding percent defective in Table 9.1.

$$\text{Distance above lower spec. limit} \ = \ \frac{\mu - \text{LSL}}{\sigma}$$

If there are both upper and lower specification limits, calculate both the percent above the upper specification limit and the percent below the lower specification limit. Then add these two quantities together. Table 9.2 gives lower defect levels in defects per million (dpm) to make them easier to read. One percent defective is equivalent to 10,000 dpm. To translate, multiply the percent defective by 10,000 to obtain the corresponding dpm.

Table 9.1: Percent Defectives Based on Normal Distribution

Distance from Spec. Limit (in σ)	Percent Defective	Distance from Spec. Limit (in σ)	Percent Defective
5.0	0.301 dpm	2.5	0.621%
4.9	0.493 dpm	2.4	0.820%
4.8	0.807 dpm	2.3	1.07%
4.7	1.31 dpm	2.2	1.39%
4.6	2.13 dpm	2.1	1.79%
4.5	3.41 dpm	2.0	2.28%
4.4	5.43 dpm	1.9	2.87%
4.3	8.55 dpm	1.8	3.59%
4.2	13.4 dpm	1.7	4.46%
4.1	20.7 dpm	1.6	5.48%
4.0	31.7 dpm	1.5	6.68%
3.9	48.1 dpm	1.4	8.08%
3.8	72.4 dpm	1.3	9.68%
3.7	108 dpm	1.2	11.5%
3.6	159 dpm	1.1	13.6%
3.5	233 dpm	1.0	15.9%
3.4	337 dpm	0.9	18.4%
3.3	483 dpm	0.8	21.2%
3.2	687 dpm	0.7	24.2%
3.1	968 dpm	0.6	27.4%
3.0	0.135%	0.5	30.9%
2.9	0.187%	0.4	34.5%
2.8	0.256%	0.3	38.2%
2.7	0.347%	0.2	42.1%
2.6	0.466%	0.1	46.0%

Example: Suppose the specification limits for fill volume are 50 and 55 ml and that $\mu=52.0$ and $\sigma=0.59$. Then:

$$\frac{\text{USL}-\mu}{\sigma} = \frac{55-52.0}{0.59} = 5.08 \quad \Rightarrow \quad 0.301 \text{ dpm}$$

$$\frac{\mu-\text{LSL}}{\sigma} = \frac{52.0-50}{0.59} = 3.39 \quad \Rightarrow \quad 337 \text{ dpm}$$

The resulting estimate of the amount of product outside the specification limits is $0.301 + 337 = 337$ dpm.

Given a set of data, one can estimate the mean and standard deviation. Estimating the mean requires calculating the average of the data points. Calculating the standard deviation is a little more complicated. The best way of calculating averages and standard deviations is to use a statistical calculator. When using such a calculator, the general procedure is:

- Put the calculator in statistic mode if necessary.

- Clear the statistic registers.

- Type each of the data points followed by the "M+" key.

- When done, press the "$\overline{X}$" key to get the average and the "σ_{n-1}" key to get the standard deviation.

The instructions for your particular calculator will contain further details. Many calculators have both "σ_n" and "σ_{n-1}" keys for calculating standard deviations. Always use the "σ_{n-1}" key.

Example: The data used to construct the histogram of fill volume in Figure 9.1 is shown in Table 9.2. The average of this data is 51.994 and its standard deviation is 0.5889247. Rounding these values gives estimates of $\mu=52.0$ and $\sigma=0.59$. Assuming an upper specification limit of 55 and a lower specification limit of 50, the defect rate was previously estimated to be 337 dpm.

Exercises

(9.1)   Calculate the mean and standard deviation of the ten numbers in the first row of Table 9.2.

(9.2)   For the filler, the specification limits are 50 and 55 ml. What is the percent defective if the standard deviation remains 0.59 but the mean decreases to 51.5?

Table 9.2: Fill Volume Data

52.0	51.7	51.7	51.3	50.8	52.6	53.0	52.5	51.9	52.2
52.4	51.3	51.7	51.8	52.0	52.0	51.8	52.0	51.5	51.5
52.1	51.5	52.2	52.2	50.9	51.4	52.9	52.7	51.6	52.7
52.6	51.2	51.6	51.0	51.7	52.3	51.8	51.9	52.6	51.8
53.0	52.0	51.9	51.8	51.7	52.9	52.5	51.2	51.6	52.3
52.1	51.9	51.4	52.4	52.6	51.8	51.8	51.4	52.8	50.8
52.3	51.7	52.6	52.5	51.8	52.6	52.5	53.7	52.7	51.8
51.8	53.1	51.4	51.2	51.8	52.0	51.9	51.8	52.4	51.3
51.7	51.3	52.5	51.4	51.4	52.4	51.8	51.3	51.7	53.2
51.9	52.9	51.1	51.6	52.7	51.5	52.0	53.3	52.0	52.5

9.2   Characterizing Variables Sampling Plans

There are two major categories of variables sampling plans: variables
sampling plans for when the standard deviation is known and
variables sampling plans for an unknown standard deviation.
Additionally, whether one has one or two specification limits effects
the selection and operation of variables sampling plans. Despite all
these differences, all variables sampling plans can be characterized
using the same two parameters:

    n = sample size

    k = accept region constant

One always starts by selecting a random sample of n units and
measuring each unit for the desired characteristic. Using these
measurements, one then calculates the sample mean. If the standard
deviation is unknown, one also calculates the sample standard
deviation. These are denoted $\overline{X}$ and S respectively.

The procedure for making accept and reject decisions depends on
whether the standard deviation is known and on the number and
type of specification limits. Table 9.3 gives the different procedures.
In Table 9.3, σ is the known standard deviation, USL is the upper
specification limit and LSL is the lower specification limit.

Example 1: Suppose the variables sampling plan n=20 and k=2.0 is
used to inspect the removal force of bottle caps. The process has a
lower specification limit of 12 pounds and a known standard
deviation of 1.2 pounds. Based on this information:

$$LSL + k\,\sigma \;=\; 12 + 2.0(1.2) \;=\; 14.4$$

The 20 samples from the most recent lot averaged 16.32 pounds.
Since this is greater than or equal to 14.4, the lot is accepted.

Example 2: Suppose the variables sampling plan n=20 and k=2.0 is used to inspect the fill weight of cereal boxes. The process has a lower specification limit of 120 grams and an upper limit of 140 grams. The standard deviation changes. It must be treated as an unknown. The 20 samples from the most recent lot averaged 124.07 grams and had a standard deviation of 2.15 grams. On the basis of this standard deviation, the acceptance limits are:

$$120 + 2.0(2.15) = 124.30 \ \leq \ \overline{X} \ \leq \ 135.70 = 140 - 2.0(2.15)$$

Since the mean does not satisfy these conditions, the lot is rejected.

Table 9.3: Procedures for Making Accept/Reject Decisions

Standard Deviation	Specification Limits	Acceptance Criteria
Known	Lower only	$LSL + k\sigma \ \leq \ \overline{X}$
	Upper only	$\overline{X} \ \leq \ USL - k\sigma$
	Upper and Lower	$LSL + k\sigma \ \leq \ \overline{X} \ \leq \ USL - k\sigma$
Unknown	Lower only	$LSL + kS \ \leq \ \overline{X}$
	Upper only	$\overline{X} \ \leq \ USL - kS$
	Upper and Lower	$LSL + kS \ \leq \ \overline{X} \ \leq \ USL - kS$

Exercises

(9.3)   Suppose the variables sampling plan n=8 and k=1.68 is used to inspect the diameter of pistons. The specification limits are 3.25 ± 0.002 inches. The standard deviation is known to be 0.00015. What are the acceptance limits for the average? Should a lot whose sample averages 3.2513 be accepted? What about a lot whose sample has an average of 3.2518?

(9.4)   Suppose the variables sampling plan n=8 and k=1.68 is used to inspect for tablet hardness. A lower limit of 4 has been established. The standard deviation changes over time, so it must be treated as an unknown. Should one accept or reject a lot whose sample has an average of 4.6 and a standard deviation of 0.23?

9.3   Evaluating Variables Sampling Plans

As with attributes sampling plans, OC curves, AOQ curves, ASN curves, AQLs, LTPDs and AOQLs are used to describe the protection provided by variables sampling plans. The program VARIABLE on the Sampling Programs diskette can be used to calculate and plot these items. Start the program as before by entering VARIABLE. Screen 9.1 shows the main menu. It has the same options for evaluating sampling plans as the other programs.

Screen 9.1: Main Menu (VARIABLE)

```
*****************************************************************************
*                           *                                             *
*                           *      Copyright (C) 1992 Taylor Enterprises  *
*    Program VARIABLE       *         P.O. Box 820, Lake Villa, IL 60046  *
*                           *                 (708) 356-1074              *
*                           *                                             *
*****************************************************************************

                  Current Sampling Plan:  no plan selected




                        -----  MENU OPTIONS  -----

            (1) Enter sampling plan
            (2) Summary information
            (3) OC curve
            (4) AOQ curve
            (5) ASN curve
            (6) Select sampling plan based on AQL and LTPD


ENTER NUMBER OF OPTION OR "q" TO QUIT  -->
```

Start by entering a sampling plan using option 1. Screen 9.2 shows the input required to enter the variables sampling plan n=26 and k=2.0 where the standard deviation is known and there are two specification limits. When asked whether to adjust for measurement variation, a reply of no was given. Measurement adjustments are covered in Section 9.9 (page 219).

Rather than ask for the standard deviation and specification limits, option 1 requests C_p. Assume the sampling plan entered is being considered for a fill volume inspection with specification limits of 50 to 55 ml. Historical data shows the standard deviation is consistent over time with a value of 0.59. C_p is then calculated as follows:

$$C_p \; = \; \frac{USL - LSL}{6\sigma} \; = \; \frac{55 - 50}{6(0.59)} \; = \; 1.41$$

C_p is commonly used as part of Statistical Process Control (SPC). For our purpose, it provides the information required by the program concerning the specification limits and known standard deviation. Once entered, the sampling plan is displayed at the top of the main menu as in Screen 9.3. C_p is only required if there are two specification limits and the standard deviation is known.

Screen 9.2: Entering Sampling Plan (VARIABLE)

```
***********************************************************************
***                                                                ***
***           VARIABLE Option 1 - Enter Variables Sampling Plan     ***
***                                                                ***
***********************************************************************

   Enter the sample size "n" and constant "k" for the desired variables
   sampling plan.  One must specify whether the standard deviation is
   known or unknown.  If the standard deviation is known, one must also
   specify 1 or 2 spec limits.  In the case of two spec limits, the
   Cp must be entered.  Optionally one can adjust for measurment variation.
   This requires entering the percentage of the total variation resulting.
   from measurement variation.  When the standard deviation is unknown,
   a single spec limit is assumed.

ENTER SAMPLE SIZE    (n)                    --> 26
ENTER CONSTANT       (k)                    --> 2
STANDARD DEVIATION KNOWN? ("y" or "n")   --> y
MEASUREMENT ADJUSTMENT? ("y" or "n")     --> n
NUMBER OF SPEC. LIMITS? (1 or 2)           --> 2
ENTER VALUE OF Cp.                          --> 1.41
```

Screen 9.3: Sampling Plan Displayed (VARIABLE)

```
***********************************************************************
*                            *                                       *
*                            *     Copyright (C) 1992 Taylor Enterprises  *
*    Program VARIABLE        *      P.O. Box 820, Lake Villa, IL 60046    *
*                            *            (708) 356-1074                  *
*                            *                                       *
***********************************************************************

         Current Sampling Plan:  n =        26,   k =        2.0000000
                                 Standard Deviation Known
                                 2 Spec Limits.        Cp =        1.41000


                         ----- MENU OPTIONS -----

              (1) Enter sampling plan
              (2) Summary information
              (3) OC curve
              (4) AOQ curve
              (5) ASN curve
              (6) Select sampling plan based on AQL and LTPD


ENTER NUMBER OF OPTION OR "q" TO QUIT  -->
```

Option 2 can now be used to obtain summary statistics. They are shown in Screen 9.4. This sampling plan has an AQL of 1.0%, LTPD of 4.0%, and AOQL of 1.24%. The program VARIABLE is slower than the other programs, especially in the case of two specification limits.

Screen 9.4: Summary Information (VARIABLE)

```
                             SUMMARY STATISTICS

         Current Sampling Plan:  n =          26,   k =        2.0000000
                                 Standard Deviation Known
                                 2 Spec Limits.        Cp =      1.41000


                                                    Process
                     Percentile      Symbol     Percent Defective
                     ___________      ______     _________________

                       95.0%          AQL             1.0099230
                       90.0%                          1.2181700
                       50.0%          IQ              2.2749780
                       10.0%          LTPD            4.0174600
                        5.0%                          4.6730580


         AOQL =        1.2433560%    (Assumes 100% inspection of rejected
                                      lots with all defectives being found
                                      and repaired plus a large lot size.)

 PRESS ENTER KEY TO CONTINUE
```

Option 3 is used to plot and tabulate OC curves. The program asks whether you want a plot or table and for information on scaling. Screen 9.5 shows a plot of the OC curve using the default scale.

Screen 9.5: OC Curve (VARIABLE)

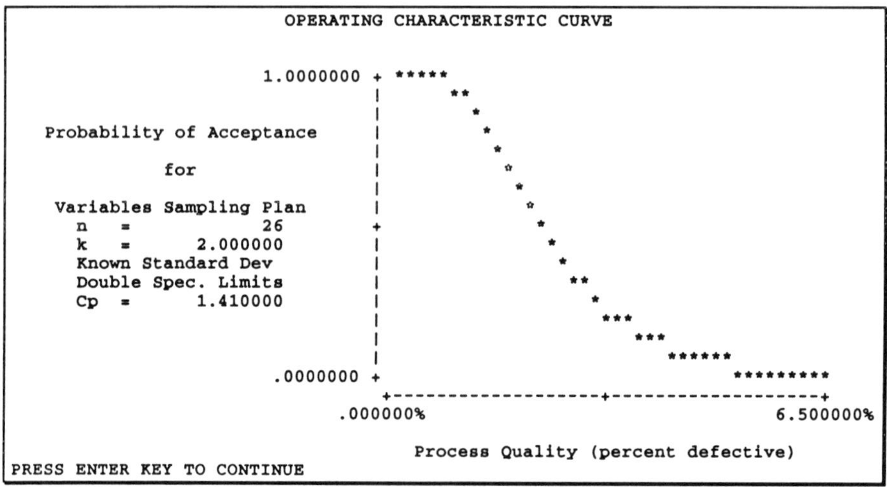

The OC curve in Screen 9.5 has the process percent defective on the bottom axis. In Section 9.1 it was shown how to use the mean and standard deviation to calculate the process percent defective. Figure 9.4 shows the effect of the mean on the process percent defective assuming $\sigma=0.59$. Combining Screen 9.5 and Figure 9.4, one can plot the probability of acceptance as a function of the mean. Figure 9.5 shows the resulting plot.

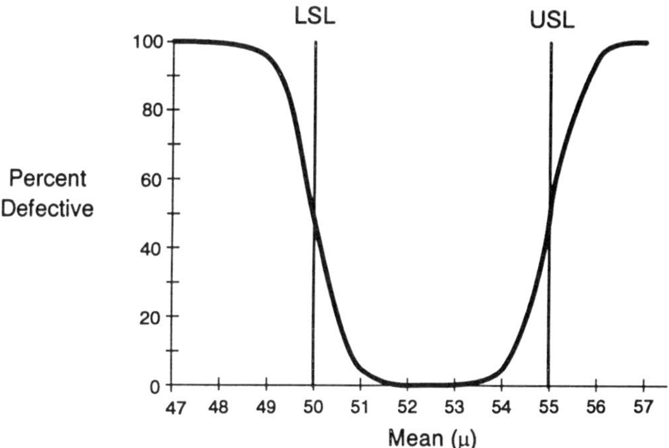

Figure 9.4: Effect of Mean on Process Quality When
LSL=50, USL=55, σ=0.59 (C_p=1.41)

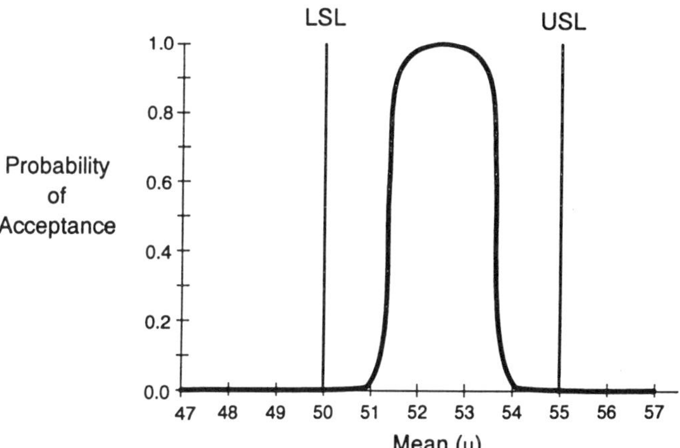

Figure 9.5: Probability of Acceptance as Function of Process Mean
When n=26, k=2.0, LSL=50, USL=55, σ=0.59

Option 4 is used to obtain plots or tables of AOQ curves. One must input the efficiency of the 100% inspection, the lot size and the method of handling defectives. The only assumption made is that rejected lots are 100% inspected. Screen 9.6 shows the AOQ curve of the variables sampling plan n=26 and k=2.0 when the 100% inspection is 100% efficient, the lots size is large and defectives are replaced. These are the assumptions used by option 2 in calculating the AOQL. The maximum value of the AOQ curve is around 1.25. This agrees with the AOQL of 1.24% calculated using option 2.

Screen 9.6: AOQ Curve (VARIABLE)

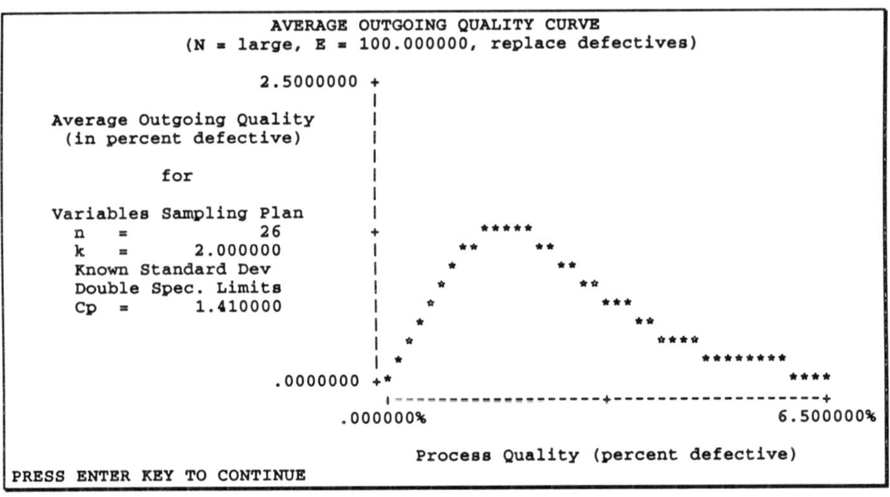

Option 5 plots and tabulates ASN curves. However, variables sampling plans have a constant sample size so the ASN is constant. All calculations done by the program VARIABLE are exact. As a result, they are frequently slow, especially when examining variables sampling plans with two specification limits.

Exercise

(9.5)   Suppose the variables sampling plan n=26 and k=2.0 is used to inspect fill volume where there is only a lower specification limit of 50 ml. Assume the standard deviation is known to be equal to 0.59. What are the AQL, LTPD and AOQL? What is the probability of accepting 2% defective lots? What is the AOQ at 2% assuming the 100% inspection is 80% effective, the lots are of 1000 units and defectives are discarded? What is the ASN at 2% defective?

9.4   Selecting Variables Sampling Plans

Variables sampling plans, like the other types of sampling plans can be selected based on their AQL and LTPD. This can be accomplished using either the program VARIABLE or Tables 9.6 and 9.7.

Tables 9.6 and 9.7 contain variables sampling plans indexed by their AQLs and LTPDs. The same AQLs and LTPDs are used as Table 3.3 for single sampling plans, Table 6.3 for double sampling plans and Table 7.2 for quick switching systems. Tables 9.6 and 9.7 assume a single specification limit. Table 9.6 is for known standard deviations and Table 9.7 is for unknown standard deviations. The two parameters given are n and k respectively. The actual AQL, LTPD and AOQL of each sampling plan are also given.

While Tables 9.6 and 9.7 assume a single specification limit, they can frequently be used for two specification limits. To determine if this is the case, calculate C_p and then compare it to the minimum value in Table 9.4. If C_p is greater than or equal to this minimum, the sampling plan can be used for two specification limits.

Table 9.4: Minimum C_p Required to Use Table 9.6 Sampling Plans For Double Specification Limits

AQL	10%	6.5%	4.0%	2.5%	1.5%	1.0%	0.65%
C_p	0.55	0.62	0.68	0.75	0.81	0.86	0.91

AQL	0.4%	0.25%	0.15%	0.1%	0.065%	0.04%	0.025%
C_p	0.96	1.01	1.06	1.10	1.14	1.18	1.22

AQL	0.015%	0.01%	0.0065%	0.004%	0.0025%
C_p	1.26	1.30	1.33	1.37	1.41

For example, the variables sampling plan n=13 and k=1.87 from Table 9.6 where the standard deviation is known. It has an AQL of 1.00% and LTPD of 6.49% assuming a single specification limit. For two specification limits, Table 9.5 shows the effect of C_p on the AQL and LTPD. Over a wide range, C_p has little effect. Only for small values of C_p are the AQL and LTPD effected. For an AQL of 1%, Table 9.4 gives a minimum C_p of 0.86. So long as the actual C_p is above this value, the actual AQL is below 1.19%.

Table 9.5: Effect of C_p on the AQL and LTPD (n=13, k= 1.87)

	2 Specification Limits						1 Limit
C_p	2	1.5	1	0.9	0.8	0.7	-
AQL	1.00	1.00	1.01	1.11	1.64	3.57	1.00
LTPD	6.49	6.49	6.49	6.50	6.55	6.83	6.49

To understand why C_p has the effect it does, look at Figure 9.4 (page 201). This figure corresponds to a C_p of 1.41. As the mean approaches either specification limit, the defect rate increases. In the middle of the chart the defect rate drops to essentially zero. The process can only make defects on the high side or the low side but not both. So long as this remains the case, a sampling plan can be selected for a single specification limit and applied to two limits.

Figure 9.6 shows the situation when C_p is 0.7. In this case, even if the process is perfectly centered, defects are made, some on the high side and some on the low side. The best the process can do is 3.57% defective. This is above the original AQL. This results in the AQL increasing to at least this minimum value.

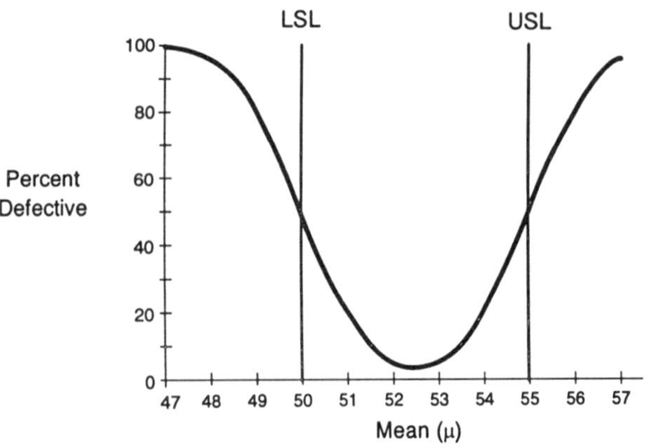

**Figure 9.6: Effect of Mean on Process Quality When
LSL=50, USL=55 and σ=1.19 (C_p=0.7)**

When the standard deviation is unknown, there is not a consistent value for C_p. It varies from lot to lot. However, the sampling plans in Table 9.7 for single specification limits can still be used. The AQL will increase significantly when the standard deviation becomes large. Despite this, the LTPD is not greatly effected.

Tables 9.8 and 9.9 (page 210) provide variables sampling plans offering the same protection as the single sampling plans in Mil-Std-105E. These tables are similar to Table 8.4 for double sampling plans and Table 8.5 for quick switching systems. Table 9.8 assumes a known standard deviation while Table 9.9 assumes an unknown standard deviation. Both tables assume a single specification limit but can frequently be applied to the case of two limits.

Option 6 of the program VARIABLE can also be used to select variables sampling plans with specified AQLs and LTPDs. One must specify the AQL, LTPD, whether the standard deviation is known, and the number of specification limits. The program is limited to known standard deviations with one or two specification limits and unknown standard deviations with a single specification limit. If a double specification limit is specified, one must also enter the C_p. Screen 9.7 shows the input required to select the variables sampling plan with an AQL of 1% and LTPD of 10% for a known standard deviation, two specification limits, and a C_p of 1.5. The selected plan is n=8 and k=1.745.

Screen 9.7: Selecting Variables Sampling Plan (VARIABLE)

```
********************************************************************************
***                                                                        ***
***      VARIABLE Option 6 - Select Sampling Plan Based on AQL and LTPD     ***
***                                                                        ***
********************************************************************************

   This option selects the variables sampling plan minimizing the sample
   size from among those satisfying the following set of conditions:

      Actual AQL is greater than or equal to the specified AQL
      Actual LTPD is less than or equal to the specified LTPD

   One must specify whether the standard deviation is known or unknown.
   If the standard deviation is known, one must also specify 1 or 2 spec
   limits.  In the case of two spec limits, the Cp must be entered.
   Optionally one can adjust for measurement variation.  This requires
   entering the percentage of the total variation resulting from
   measurement variation.  When the standard deviation is unknown,
   a single spec limit is assumed.

ENTER AQL  AS PERCENT DEFECTIVE           --> 1
ENTER LTPD AS PERCENT DEFECTIVE           --> 10
STANDARD DEVIATION KNOWN? ("y" or "n")    --> y
MEASUREMENT ADJUSTMENT? ("y" or "n")      --> n
NUMBER OF SPEC. LIMITS? (1 or 2)          --> 2
ENTER VALUE OF Cp.                        --> 1.5
USE .95 & .1 FOR AQL-LTPD? ("y" or "n")   --> y
```

Table 9.6: Variables Sampling Plans - Standard Deviation Known
(Single Specification Limit)

AQL	Approximate Ratio of $LTPD/AQL$								
	45	**11**	**6.5**	**5**	**4**	**3.2**	**2.8**	**2.3**	**2**
10%	-	**3, 0.33** AQL = 10.0 LTPD = 65.9 AOQL = 19.0	**6, 0.61** AQL = 10.0 LTPD = 46.5 AOQL = 14.4	**8, 0.70** AQL = 10.0 LTPD = 40.3 AOQL = 13.2	**10, 0.76** AQL = 10.0 LTPD = 36.1 AOQL = 12.4	**15, 0.85** AQL = 10.1 LTPD = 30.2 AOQL = 11.5	**24, 0.94** AQL = 10.1 LTPD = 24.9 AOQL = 10.7	**36, 1.00** AQL = 10.1 LTPD = 21.6 AOQL = 10.2	**50, 1.05** AQL = 9.98 LTPD = 19.2 AOQL = 9.86
6.5%	-	**4, 0.69** AQL = 6.52 LTPD = 48.0 AOQL = 12.4	**7, 0.89** AQL = 6.53 LTPD = 34.3 AOQL = 9.76	**9, 0.96** AQL = 6.57 LTPD = 29.7 AOQL = 8.99	**14, 1.07** AQL = 6.56 LTPD = 23.3 AOQL = 7.93	**20, 1.14** AQL = 6.58 LTPD = 19.7 AOQL = 7.39	**34, 1.23** AQL = 6.52 LTPD = 15.6 AOQL = 6.79	**46, 1.27** AQL = 6.52 LTPD = 14.0 AOQL = 6.58	**70, 1.32** AQL = 6.47 LTPD = 12.2 AOQL = 6.35
4.0%	-	**5, 1.01** AQL = 4.04 LTPD = 33.1 AOQL = 7.88	**8, 1.17** AQL = 3.99 LTPD = 23.7 AOQL = 6.24	**10, 1.23** AQL = 4.00 LTPD = 20.5 AOQL = 5.74	**16, 1.34** AQL = 3.99 LTPD = 15.4 AOQL = 4.95	**22, 1.40** AQL = 4.00 LTPD = 13.0 AOQL = 4.60	**38, 1.48** AQL = 4.03 LTPD = 10.2 AOQL = 4.24	**55, 1.53** AQL = 3.99 LTPD = 8.74 AOQL = 4.04	**80, 1.57** AQL = 3.97 LTPD = 7.68 AOQL = 3.91
2.5%	**2, 0.80** AQL = 2.48 LTPD = 54.2 AOQL = 10.7	**5, 1.22** AQL = 2.52 LTPD = 25.9 AOQL = 5.57	**10, 1.44** AQL = 2.50 LTPD = 15.0 AOQL = 3.86	**12, 1.49** AQL = 2.47 LTPD = 13.1 AOQL = 3.56	**20, 1.59** AQL = 2.51 LTPD = 9.62 AOQL = 3.08	**26, 1.64** AQL = 2.48 LTPD = 8.25 AOQL = 2.87	**45, 1.71** AQL = 2.53 LTPD = 6.44 AOQL = 2.66	**62, 1.75** AQL = 2.51 LTPD = 5.62 AOQL = 2.55	**90, 1.79** AQL = 2.48 LTPD = 4.90 AOQL = 2.45
1.5%	**2, 1.00** AQL = 1.53 LTPD = 46.3 AOQL = 8.23	**6, 1.50** AQL = 1.49 LTPD = 16.4 AOQL = 3.34	**10, 1.65** AQL = 1.50 LTPD = 10.7 AOQL = 2.51	**15, 1.74** AQL = 1.52 LTPD = 7.94 AOQL = 2.14	**22, 1.82** AQL = 1.50 LTPD = 6.10 AOQL = 1.87	**30, 1.87** AQL = 1.50 LTPD = 5.09 AOQL = 1.74	**46, 1.93** AQL = 1.49 LTPD = 4.08 AOQL = 1.60	**65, 1.97** AQL = 1.49 LTPD = 3.51 AOQL = 1.53	**95, 2.00** AQL = 1.50 LTPD = 3.08 AOQL = 1.50
1.0%	**2, 1.17** AQL = 0.982 LTPD = 39.6 AOQL = 6.47	**7, 1.70** AQL = 1.01 LTPD = 11.2 AOQL = 2.23	**13, 1.87** AQL = 1.00 LTPD = 6.49 AOQL = 1.57	**18, 1.93** AQL = 1.02 LTPD = 5.18 AOQL = 1.41	**26, 2.00** AQL = 1.01 LTPD = 4.02 AOQL = 1.24	**36, 2.05** AQL = 1.01 LTPD = 3.32 AOQL = 1.15	**60, 2.11** AQL = 1.01 LTPD = 2.59 AOQL = 1.06	**88, 2.15** AQL = 1.00 LTPD = 2.20 AOQL = 1.01	**120, 2.17** AQL = 1.02 LTPD = 2.00 AOQL = 1.00
0.65%	**3, 1.53** AQL = 0.657 LTPD = 21.5 AOQL = 3.35	**8, 1.90** AQL = 0.654 LTPD = 7.40 AOQL = 1.44	**15, 2.06** AQL = 0.648 LTPD = 4.19 AOQL = 1.01	**20, 2.12** AQL = 0.643 LTPD = 3.34 AOQL = 0.891	**32, 2.19** AQL = 0.655 LTPD = 2.48 AOQL = 0.787	**42, 2.23** AQL = 0.650 LTPD = 2.11 AOQL = 0.736	**72, 2.29** AQL = 0.650 LTPD = 1.62 AOQL = 0.677	**100, 2.32** AQL = 0.649 LTPD = 1.42 AOQL = 0.654	**150, 2.35** AQL = 0.649 LTPD = 1.24 AOQL = 0.636
0.4%	**3, 1.71** AQL = 0.391 LTPD = 16.6 AOQL = 2.40	**9, 2.10** AQL = 0.404 LTPD = 4.72 AOQL = 0.895	**16, 2.25** AQL = 0.389 LTPD = 2.68 AOQL = 0.622	**22, 2.30** AQL = 0.402 LTPD = 2.13 AOQL = 0.560	**34, 2.37** AQL = 0.400 LTPD = 1.58 AOQL = 0.487	**46, 2.41** AQL = 0.399 LTPD = 1.32 AOQL = 0.454	**74, 2.46** AQL = 0.401 LTPD = 1.04 AOQL = 0.422	**110, 2.50** AQL = 0.394 LTPD = 0.871 AOQL = 0.398	-
0.25%	**4, 1.98** AQL = 0.253 LTPD = 9.03 AOQL = 1.24	**10, 2.28** AQL = 0.255 LTPD = 3.04 AOQL = 0.566	**18, 2.42** AQL = 0.249 LTPD = 1.71 AOQL = 0.395	**25, 2.47** AQL = 0.256 LTPD = 1.34 AOQL = 0.353	**38, 2.54** AQL = 0.250 LTPD = 0.985 AOQL = 0.304	**50, 2.57** AQL = 0.253 LTPD = 0.845 AOQL = 0.289	**82, 2.63** AQL = 0.246 LTPD = 0.641 AOQL = 0.259	-	-
0.15%	**4, 2.14** AQL = 0.153 LTPD = 6.69 AOQL = 0.907	**11, 2.47** AQL = 0.151 LTPD = 1.86 AOQL = 0.339	**20, 2.60** AQL = 0.150 LTPD = 1.04 AOQL = 0.237	**26, 2.65** AQL = 0.148 LTPD = 0.823 AOQL = 0.210	**38, 2.70** AQL = 0.150 LTPD = 0.635 AOQL = 0.188	**52, 2.74** AQL = 0.150 LTPD = 0.520 AOQL = 0.173	-	-	-

Table 9.6: (continued)

AQL	Approximate Ratio of $LTPD/AQL$								
	45	**11**	**6.5**	**5**	**4**	**3.2**	**2.8**	**2.3**	**2**
0.1%	5, 2.35 AQL = 0.102 LTPD = 3.78 AOQL = 0.517	12, 2.61 AQL = 0.102 LTPD = 1.25 AOQL = 0.227	24, 2.75 AQL = 0.101 LTPD = 0.642 AOQL = 0.153	30, 2.79 AQL = 0.100 LTPD = 0.529 AOQL = 0.138	45, 2.84 AQL = 0.102 LTPD = 0.404 AOQL = 0.123	·			·
0.065%	5, 2.48 AQL = 0.0651 LTPD = 2.83 AOQL = 0.371	14, 2.77 AQL = 0.0664 LTPD = 0.760 AOQL = 0.140	26, 2.89 AQL = 0.0658 LTPD = 0.416 AOQL = 0.0987	34, 2.93 AQL = 0.0659 LTPD = 0.336 AOQL = 0.0890	·	·	·	·	·
0.04%	6, 2.68 AQL = 0.0402 LTPD = 1.55 AOQL = 0.205	15, 2.92 AQL = 0.0412 LTPD = 0.481 AOQL = 0.0877	26, 3.03 AQL = 0.0400 LTPD = 0.273 AOQL = 0.0622	·	·	·	·	·	·
0.025%	6, 2.81 AQL = 0.0249 LTPD = 1.11 AOQL = 0.142	16, 3.07 AQL = 0.0250 LTPD = 0.298 AOQL = 0.0536	·	·	·	·	·	·	·
0.015%	7, 2.99 AQL = 0.0152 LTPD = 0.611 AOQL = 0.0787	16, 3.20 AQL = 0.0152 LTPD = 0.199 AOQL = 0.0345	·	·	·	·	·	·	·
0.01%	7, 3.09 AQL = 0.0103 LTPD = 0.459 AOQL = 0.0576	·	·	·	·	·	·	·	·
0.0065%	8, 3.24 AQL = 0.00663 LTPD = 0.266 AOQL = 0.0338	·	·	·	·	·	·	·	·
0.004%	8, 3.36 AQL = 0.00405 LTPD = 0.183 AOQL = 0.0226	·	·	·	·	·	·	·	·
0.0025%	9, 3.50 AQL = 0.00258 LTPD = 0.106 AOQL = 0.0133	·	·	·	·	·	·	·	·

Table 9.7: Variables Sampling Plans - Standard Deviation Unknown
(Single Specification Limit)

AQL	Approximate Ratio of $LTPD/AQL$								
	45	**11**	**6.5**	**5**	**4**	**3.2**	**2.8**	**2.3**	**2**
10%	-	**3, 0.33** AQL = 10.1 LTPD = 68.2 AOQL = 19.7	**6, 0.58** AQL = 9.90 LTPD = 50.7 AOQL = 15.3	**10, 0.71** AQL = 10.0 LTPD = 40.7 AOQL = 13.3	**14, 0.78** AQL = 10.2 LTPD = 35.5 AOQL = 12.4	**20, 0.86** AQL = 9.97 LTPD = 30.4 AOQL = 11.4	**36, 0.95** AQL = 10.1 LTPD = 24.6 AOQL = 10.6	**55, 1.01** AQL = 10.0 LTPD = 21.4 AOQL = 10.2	**80, 1.05** AQL = 10.1 LTPD = 19.3 AOQL = 9.95
6.5%	-	**5, 0.71** AQL = 6.55 LTPD = 49.3 AOQL = 12.9	**10, 0.91** AQL = 6.49 LTPD = 34.5 AOQL = 9.86	**13, 0.97** AQL = 6.52 LTPD = 30.3 AOQL = 9.13	**22, 1.08** AQL = 6.50 LTPD = 23.5 AOQL = 7.96	**32, 1.14** AQL = 6.58 LTPD = 20.1 AOQL = 7.47	**60, 1.23** AQL = 6.58 LTPD = 15.7 AOQL = 6.85	**85, 1.27** AQL = 6.59 LTPD = 14.0 AOQL = 6.65	**140, 1.32** AQL = 6.45 LTPD = 12.1 AOQL = 6.40
4.0%	-	**7, 1.00** AQL = 4.06 LTPD = 36.0 AOQL = 8.55	**14, 1.19** AQL = 3.97 LTPD = 23.7 AOQL = 6.30	**20, 1.26** AQL = 4.06 LTPD = 19.5 AOQL = 5.68	**32, 1.35** AQL = 4.04 LTPD = 15.3 AOQL = 4.99	**46, 1.41** AQL = 4.02 LTPD = 12.8 AOQL = 4.61	**80, 1.48** AQL = 4.07 LTPD = 10.2 AOQL = 4.29	**130, 1.53** AQL = 3.99 LTPD = 8.63 AOQL = 4.10	**180, 1.56** AQL = 4.01 LTPD = 7.82 AOQL = 4.01
2.5%	**2, 0.71** AQL = 2.51 LTPD = 68.3 AOQL = 14.8	**9, 1.24** AQL = 2.53 LTPD = 26.5 AOQL = 5.80	**20, 1.44** AQL = 2.52 LTPD = 15.6 AOQL = 4.01	**28, 1.51** AQL = 2.51 LTPD = 12.7 AOQL = 3.56	**45, 1.60** AQL = 2.47 LTPD = 9.65 AOQL = 3.07	**62, 1.64** AQL = 2.52 LTPD = 8.32 AOQL = 2.92	**120, 1.71** AQL = 2.51 LTPD = 6.36 AOQL = 2.72	**170, 1.75** AQL = 2.51 LTPD = 5.56 AOQL = 2.59	**250, 1.79** AQL = 2.48 LTPD = 4.86 AOQL = 2.47
1.5%	**3, 1.02** AQL = 1.52 LTPD = 51.5 AOQL = 9.73	**12, 1.49** AQL = 1.51 LTPD = 18.0 AOQL = 3.68	**26, 1.68** AQL = 1.49 LTPD = 10.2 AOQL = 2.47	**38, 1.74** AQL = 1.55 LTPD = 8.11 AOQL = 2.21	**60, 1.82** AQL = 1.53 LTPD = 6.15 AOQL = 1.91	**80, 1.87** AQL = 1.49 LTPD = 5.19 AOQL = 1.75	**140, 1.92** AQL = 1.52 LTPD = 4.13 AOQL = 1.68	**200, 1.96** AQL = 1.51 LTPD = 3.56 AOQL = 1.59	**300, 2.00** AQL = 1.50 LTPD = 3.06 AOQL = 1.51
1.0%	**5, 1.33** AQL = 1.00 LTPD = 32.8 AOQL = 5.66	**18, 1.72** AQL = 1.01 LTPD = 11.2 AOQL = 2.28	**38, 1.88** AQL = 1.02 LTPD = 6.42 AOQL = 1.60	**50, 1.94** AQL = 0.990 LTPD = 5.24 AOQL = 1.41	**80, 2.01** AQL = 1.00 LTPD = 3.97 AOQL = 1.24	**120, 2.04** AQL = 1.02 LTPD = 3.34 AOQL = 1.21	**200, 2.11** AQL = 0.998 LTPD = 2.58 AOQL = 1.08	**300, 2.15** AQL = 0.996 LTPD = 2.20 AOQL = 1.02	**410, 2.17** AQL = 1.01 LTPD = 2.00 AOQL = 1.01
0.65%	**6, 1.51** AQL = 0.657 LTPD = 25.6 AOQL = 4.14	**24, 1.9** AQL = 0.659 LTPD = 7.27 AOQL = 1.46	**48, 2.07** AQL = 0.646 LTPD = 4.18 AOQL = 1.02	**68, 2.13** AQL = 0.644 LTPD = 3.28 AOQL = 0.893	**110, 2.18** AQL = 0.647 LTPD = 2.54 AOQL = 0.834	**160, 2.23** AQL = 0.649 LTPD = 2.07 AOQL = 0.758	**275, 2.29** AQL = 0.648 LTPD = 1.61 AOQL = 0.687	**380, 2.32** AQL = 0.645 LTPD = 1.41 AOQL = 0.658	**560, 2.35** AQL = 0.643 LTPD = 1.24 AOQL = 0.636
0.4%	**9, 1.78** AQL = 0.401 LTPD = 15.2 AOQL = 2.33	**30, 2.11** AQL = 0.411 LTPD = 4.74 AOQL = 0.922	**60, 2.26** AQL = 0.395 LTPD = 2.63 AOQL = 0.627	**84, 2.31** AQL = 0.402 LTPD = 2.09 AOQL = 0.560	**140, 2.37** AQL = 0.396 LTPD = 1.55 AOQL = 0.505	**190, 2.41** AQL = 0.396 LTPD = 1.30 AOQL = 0.466	**310, 2.46** AQL = 0.399 LTPD = 1.03 AOQL = 0.428	**460, 2.49** AQL = 0.404 LTPD = 0.893 AOQL = 0.412	-
0.25%	**11, 1.96** AQL = 0.257 LTPD = 10.7 AOQL = 1.57	**38, 2.30** AQL = 0.253 LTPD = 2.96 AOQL = 0.565	**75, 2.43** AQL = 0.252 LTPD = 1.67 AOQL = 0.397	**110, 2.47** AQL = 0.251 LTPD = 1.31 AOQL = 0.372	**170, 2.53** AQL = 0.255 LTPD = 0.997 AOQL = 0.323	**225, 2.57** AQL = 0.251 LTPD = 0.838 AOQL = 0.295	**380, 2.62** AQL = 0.254 LTPD = 0.655 AOQL = 0.271	-	-
0.15%	**14, 2.17** AQL = 0.151 LTPD = 6.74 AOQL = 0.948	**44, 2.47** AQL = 0.152 LTPD = 1.94 AOQL = 0.356	**90, 2.60** AQL = 0.154 LTPD = 1.04 AOQL = 0.244	**125, 2.64** AQL = 0.149 LTPD = 0.829 AOQL = 0.226	**190, 2.70** AQL = 0.150 LTPD = 0.623 AOQL = 0.194	**250, 2.74** AQL = 0.146 LTPD = 0.519 AOQL = 0.176	-	-	-

Table 9.7: (continued)

AQL	Approximate Ratio of $LTPD/AQL$								
	45	11	6.5	5	4	3.2	2.8	2.3	2
0.1%	18, 2.34 AQL = 0.103 LTPD = 4.29 AOQL = 0.598	55, 2.62 AQL = 0.103 LTPD = 1.24 AOQL = 0.230	120, 2.74 AQL = 0.101 LTPD = 0.651 AOQL = 0.166	150, 2.78 AQL = 0.0995 LTPD = 0.541 AOQL = 0.148	240, 2.84 AQL = 0.101 LTPD = 0.397 AOQL = 0.127	.	.	.	.
0.065%	22, 2.51 AQL = 0.0647 LTPD=2.74 AOQL=0.374	70, 2.78 AQL = 0.0664 LTPD = 0.751 AOQL = 0.142	140, 2.88 AQL = 0.0657 LTPD = 0.423 AOQL = 0.107	190, 2.93 AQL = 0.0648 LTPD = 0.331 AOQL = 0.0925	.	.	.	.	.
0.04%	26, 2.67 AQL = 0.0401 LTPD = 1.77 AOQL = 0.236	80, 2.93 AQL = 0.0405 LTPD = 0.478 AOQL = 0.0882	160, 3.03 AQL = 0.0400 LTPD = 0.264 AOQL = 0.0656	.	.	.	.	.	.
0.025%	30, 2.82 AQL = 0.0247 LTPD = 1.15 AOQL = 0.151	90, 3.07 AQL = 0.0250 LTPD = 0.309 AOQL = 0.0557	.	.	.	.	.	.	.
0.015%	35, 2.97 AQL = 0.0153 LTPD = 0.728 AOQL = 0.0935	110, 3.20 AQL = 0.0151 LTPD = 0.189 AOQL = 0.0371	.	.	.	.	.	.	.
0.01%	42, 3.11 AQL = 0.0101 LTPD = 0.449 AOQL = 0.0578	.	.	.	.	.	.	.	.
0.0065%	50, 3.25 AQL = 0.00646 LTPD = 0.273 AOQL = 0.0352	.	.	.	.	.	.	.	.
0.004%	55, 3.37 AQL = 0.00410 LTPD = 0.182 AOQL = 0.0231	.	.	.	.	.	.	.	.
0.0025%	62, 3.50 AQL = 0.00253 LTPD = 0.114 AOQL = 0.0143	.	.	.	.	.	.	.	.

Table 9.8: Matching Variables Sampling Plans for Mil-Std-105E - Standard Deviation Known

Sample Size	Acceptance Number										
	0	1	2	3	5	7	10	14	21	30	44
3	2, 0.96	*	*	*	*	*	*	*	*	*	*
5	2, 1.17	3, 0.52	*	*	*	*	*	*	*	*	*
8	3, 1.54	4, 0.86	5, 0.48	*	*	*	*	*	*	*	*
13	3, 1.72	5, 1.18	7, 0.88	8, 0.63	*	*	*	*	*	*	*
20	4, 1.98	6, 1.42	8, 1.14	10, 0.94	12, 0.61	*	*	*	*	-	-
32	4, 2.12	7, 1.67	10, 1.42	12, 1.23	15, 0.94	18, 0.72	*	*	*	-	-
50	5, 2.34	8, 1.87	12, 1.65	14, 1.48	18, 1.23	22, 1.04	25, 0.80	*	*	-	-
80	5, 2.48	10, 2.09	14, 1.87	16, 1.71	22, 1.48	28, 1.32	32, 1.12	38, 0.91	*	-	-
125	6, 2.67	11, 2.26	16, 2.06	20, 1.92	26, 1.71	32, 1.55	40, 1.38	50, 1.20	60, 0.95	-	-
200	6, 2.80	12, 2.44	18, 2.25	22, 2.11	30, 1.92	36, 1.78	46, 1.62	56, 1.46	70, 1.24	-	-
315	7, 2.97	14, 2.61	20, 2.42	25, 2.29	34, 2.11	42, 1.98	54, 1.84	70, 1.68	85, 1.49	-	-
500	7, 3.09	15, 2.76	22, 2.59	28, 2.46	40, 2.30	50, 2.17	60, 2.04	80, 1.90	100, 1.72	-	-
800	8, 3.24	16, 2.91	24, 2.74	30, 2.63	44, 2.47	54, 2.35	70, 2.22	90, 2.10	115, 1.93	-	-
1250	8, 3.36	18, 3.05	26, 2.89	34, 2.78	50, 2.63	60, 2.51	80, 2.39	100, 2.27	130, 2.12	-	-
2000	-	20, 3.20	28, 3.03	36, 2.93	52, 2.78	70, 2.68	90, 2.56	110, 2.44	150, 2.30	-	-

*   The single sampling plan in Mil-Std-105E is intended for tallying defects, not defectives.

Table 9.9: Matching Variables Sampling Plans for Mil-Std-105E - Standard Deviation Unknown

Sample Size	Acceptance Number										
	0	1	2	3	5	7	10	14	21	30	44
3	3, 0.99	*	*	*	*	*	*	*	*	*	*
5	4, 1.24	4, 0.57	*	*	*	*	*	*	*	*	*
8	6, 1.52	6, 0.91	6, 0.52	*	*	*	*	*	*	*	*
13	9, 1.78	9, 1.21	9, 0.88	9, 0.63	*	*	*	*	*	*	*
20	11, 1.96	13, 1.45	14, 1.16	14, 0.94	14, 0.61	*	*	*	*	-	-
32	14, 2.15	18, 1.69	20, 1.43	22, 1.24	22, 0.95	22, 0.73	*	*	*	-	-
50	18, 2.34	24, 1.89	28, 1.66	30, 1.48	34, 1.23	34, 1.04	34, 0.81	*	*	-	-
80	22, 2.51	32, 2.10	38, 1.87	42, 1.72	48, 1.49	50, 1.32	52, 1.12	54, 0.91	*	-	-
125	26, 2.66	40, 2.27	50, 2.06	55, 1.92	65, 1.71	70, 1.56	75, 1.38	80, 1.20	85, 0.95	-	-
200	30, 2.82	50, 2.45	60, 2.25	70, 2.12	85, 1.93	100, 1.79	115, 1.62	125, 1.46	130, 1.23	-	-
315	36, 2.96	60, 2.61	75, 2.42	90, 2.29	115, 2.10	130, 1.98	150, 1.83	170, 1.68	190, 1.49	-	-
500	42, 3.10	70, 2.76	95, 2.59	120, 2.46	150, 2.29	170, 2.17	200, 2.03	225, 1.89	250, 1.72	-	-
800	50, 3.24	85, 2.92	120, 2.73	140, 2.62	180, 2.46	210, 2.35	250, 2.22	290, 2.10	340, 1.93	-	-
1250	55, 3.37	100, 3.05	140, 2.88	170, 2.77	220, 2.62	260, 2.51	310, 2.39	360, 2.27	440, 2.11	-	-
2000	-	120, 3.18	160, 3.03	200, 2.92	260, 2.78	315, 2.68	380, 2.56	450, 2.44	550, 2.30	-	-

* The single sampling plan in Mil-Std-105E is intended for tallying defects, not defectives.

Exercises

(9.6)   Select a variables sampling plan with an AQL of 1% and LTPD of 10% when the standard deviation is known and there is only one specification limit. How does this compare to the sampling plan selected in Screen 9.7? Assume the lower specification limit is 25.0 and the standard deviation is 2.2. What is the acceptance limit for this sampling plan?

(9.7)   Select a variables sampling plan with an AQL of 1% and LTPD of 10% when the standard deviation is unknown and there is only one specification limit. How does this compare to the sampling plan selected in exercise 9.6? Assume the lower specification limit is 25.0. Should a lot with a mean of 26.57 and standard deviation of 1.24 be accepted or rejected?

9.5   Known Standard Deviation

Assuming that the standard deviation is known requires that the process variation be consistent over time. This assumption can be verified by maintaining a control chart of the standard deviation. It also requires an accurate estimate of the standard deviation. Obtaining such an estimate requires a minimum of 100 samples. An estimate of the standard deviation based on 100 samples is accurate to within plus or minus 15%. Even with this accuracy, the resulting errors in the estimated percent defective can be sizable. The decision to use a known standard deviation plan should not be treated lightly.

If a known standard deviation plan is used:

- The standard deviation should first be control charted to demonstrate that it is constant. This control chart should be maintained for several weeks.

- An estimate of the standard deviation should be made based on at least 100 samples. Several hundred samples are preferable.

- Once the sampling plan is implemented, a separate control chart of the standard deviation should be maintained. If there is evidence of a change in the standard deviation, the known standard deviation plan should not be used.

A variables sampling plan with known standard deviation can be combined with a control chart to form an acceptance-control chart[1]. Figure 9.7 shows an example of such a chart. This chart shows the average for each lot inspected along with four decision lines. The inner two lines, UCL_X and LCL_X, are called the upper and lower control limits respectively. A point outside these two lines indicates that the process average has shifted. The outer two lines, UAL_X and LAL_X, are called the upper and lower acceptance limits. A point outside these two limits indicates the lot should be rejected.

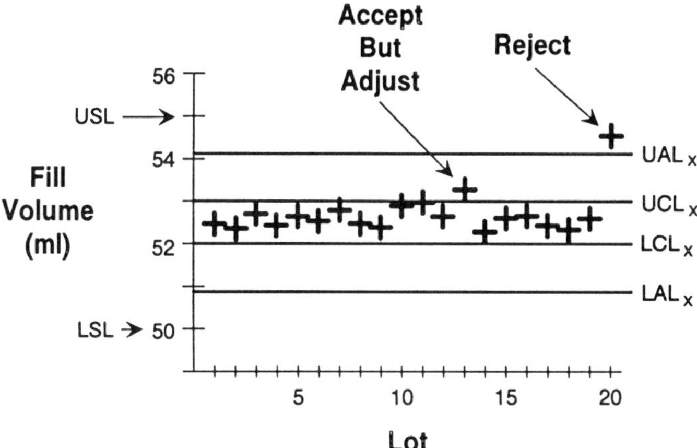

Figure 9.7: Acceptance-Control Chart

The formulas for the control and acceptance limits are as follows:

$$UAL_X = USL - k \times \sigma$$

$$UCL_X = T + 3 \, \sigma \Big/ \sqrt{n}$$

$$LCL_X = T - 3 \, \sigma \Big/ \sqrt{n}$$

$$LAL_X = LSL + k \times \sigma$$

Here, n and k are the parameters of the variables sampling plan, σ is the known standard deviation and T is the target. Frequently T is half way between the specification limits, i.e., T=(USL+LSL)/2. Note that accepting a point that is between or on the acceptance lines is equivalent to passing the acceptance criteria given in Table 9.3. A control chart of the standard deviation should also be maintained.

---

[1]See Freund (1959)

There is a difference in the way samples are selected for control charts and sampling plans. When acceptance sampling, samples should be drawn randomly across the entire period or lot. When control charting, the sample is generally taken from the most recent production. For example, the sample might be drawn randomly from the most recent tote pan. If one is using the same samples for both control charting and acceptance sampling, how does one resolve this conflict? One procedure is to select the samples as you would for control charting but add an extra inspection before the first lot. Then for a lot to be accepted, the sampling plans at both the beginning of the lot and end of the lot must reach the accept decisions. Whenever a lot is rejected, the process is fixed or adjusted and a new sample selected. If this sample passes, one can start making releasable production again. Use of this procedure requires that the probability of two changes in the process average during a single lot be small.

It is important to understand the difference between control limits and acceptance limits. Control limits are calculated out from the target. They are used to decide if the process has shifted from target. Control limits are not dependent on the specification limits. They indicate any change, not just changes resulting in defective product. Acceptance limits on the other hand are calculated in from the specification limits. They indicate that the process average has come sufficiently close to the specification limits to cause defective product to be made. Many times, the control limits can detect a shift and the problem be corrected without the production of defective product.

A practice used by some is to reject product coming from lots or periods whose point falls outside the control limits. The result of this practice can be the rejection of large quantities of good product. While a point inside the control limits signifies good product, a point outside the control limits does not necessarily signify bad product. Separate acceptance limits should be used.

9.6　Attributes Versus Variables Sampling Plans

Attributes sampling plans make accept and reject decisions based on a tally of the number of defects or defectives. Single sampling plans, double sampling plans, and quick switching systems are examples of attributes sampling plans. With attributes sampling plans, one can combine several types of defects into a single inspection. Variables sampling plans, in contrast, make accept and reject decisions based on actual measurements. A separate sampling plan must be applied to each characteristic. It is assumed that each characteristic only occurs once per unit. OC curves of variables sampling plans are always in terms of percent defective. As a result, many of the Mil-Std-105E single sampling plans in Tables 9.8 and 9.9 do not have matching variables sampling plans. Those sampling plans not matched are intended for use when the number of defects is tallied.

Variables sampling plans can significantly reduce the sample size. However, not all pass/fail inspections can be converted to measurements. Take, for example, missing bottle caps. Bottle caps could be missing as a result of low removal forces. Specification limits for removal force can be determined and a variables sampling plan implemented. Alternatively, bottle caps could be missing as a result of operator error. No conversion is possible in this case.

Determining whether conversion is possible requires understanding the underlying cause of the defect. If the cause is a mistake in the form of an error or omission, conversion is not possible. This is the case when the operator fails to put a cap on. If the defect is effected by numerous factors, each of which must stay within prescribed limits, conversion is possible.   This is the case when the removal forces are low. Removal forces are effected by the inside diameter of the cap, the outside diameter of the neck of the bottle, the insertion force, the insertion depth and numerous other factors.

9.7   Normality Assumption

Variables sampling plans assume that the measurements follow the normal distribution shown in Figures 9.1 and 9.2 (page 192). Most statistical software packages include a procedure for testing for normality. Anderson (1974) shows how to use the W-test to test for normality. If the distribution is not normal, it may be possible to transform the data so that the transformed data is normally distributed. For example, the logarithms of the original data might be normally distributed.

Another alternative is to use variables data to screen the lots and accept those lots that are obviously good. All other lots are then inspected using an attributes sampling plan. For example, a control chart could be used where lots inside the control limits are released and lots outside the control limits are inspected using an attributes sampling plan. This procedure achieves much of the savings resulting from using measurements without such rigid assumptions. In setting up the control chart, care should be taken to ensure the control chart is capable of detecting shifts in the process average before defective product is made. This approach requires that the process be more than capable of holding the specification limits. Typically C_p should be 1.5 or greater.

Even when lots are normally distributed, one can still encouter problems. Figure 9.8 shows what can happen when the process shifts during the middle of a lot. Even though the measurements are normally distributed before and after the shift, the histogram of the entire lot is far from normal. The dotted normal curve shows the best fit normal curve for this data. Fortunately, this best fit curve over estimates the percent defective. The impact is that, while good lots may be rejected, at least bad lots are not released.

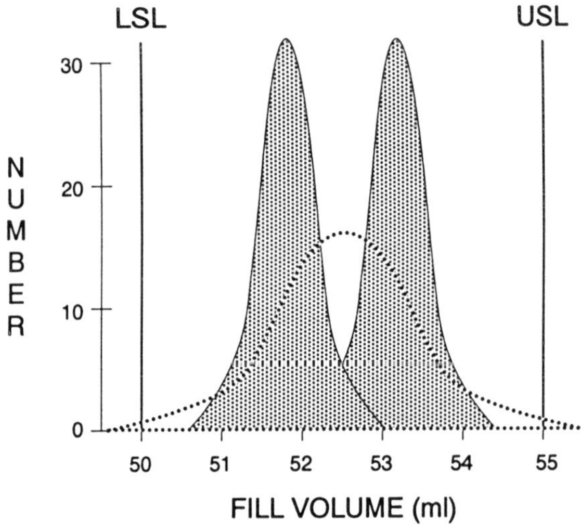

Figure 9.8: Effect of Shift in the Middle of a Lot

Variables sampling plans fail to give adequate protection against short periods of highly defective product. Figure 9.9 shows this situation. Such periods most commonly occur at the beginning or end of a lot. Suppose the variables sampling plan n=5 and k=2.35 is used to inspect this lot. This sampling plan has an AQL of 0.1% and LTPD of 3.78% when the standard deviation is known. If a unit from the highly defective period is included in the sample of 5, this unit causes the sample average to fall outside the acceptance limits. This results in the lot being rejected. Likewise, if none of the units from the highly defective period is included in the sample, the lot is accepted. This sampling plan behaves just like the single sampling plan n=5 and a=0. One has five chances of finding one of the defective units resulting in rejection. This sampling plan has an AQL of 1% and LTPD of 37%. If these highly defective periods always occur at the beginning or end of a lot, stratified samples should be selected. The first and last units of the lot should be included in the sample. Sampling at the beginning and end of a lot as explained in Section 9.5 also corrects this problem. If periods of highly defective units can occur in the middle of a lot, variables sampling should not be used.

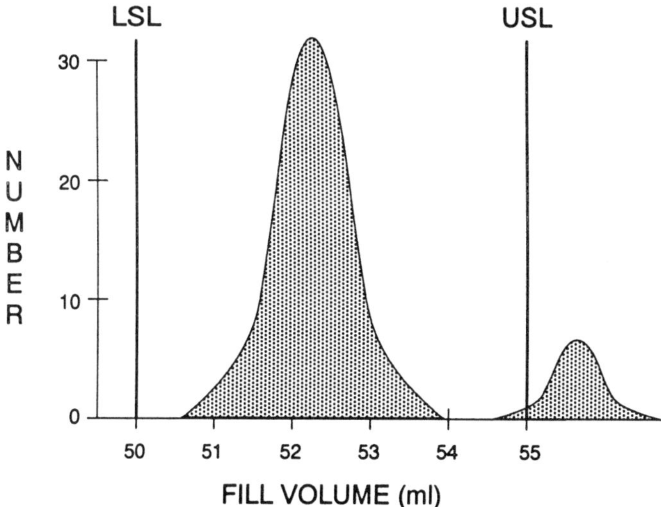

**Figure 9.9: Worse Case Scenario -
Short Period of Highly Defective Product**

Variables sampling plans also fail to perform well in the presence of outliers. For example, one might be inspecting a filling operation for fill volume. Most of the fill volume variation is a result of the filling machine. However, on an infrequent basis, a bag gets placed on the nozzle too late, is removed too early, or is left on for two cycles. This results in outliers in the form of gross underfills and overfills. Inclusion of one of these outliers in a sample always results in rejection. In protecting against such outliers, a variables sampling plan once again behaves like an attributes sampling plan with accept number of zero. In the presence of outliers, variables sampling plans should not be used on their own. If a variables sampling plan is used, a separate inspection for outliers should be performed. For example, a weight checker might be installed on-line that can identify gross overfills and underfills.

9.8   Variables Inspection for Parameters

The variables sampling plans in this chapter control the percent defective. These sampling plans react to the number of units falling above the upper specification limit or below the lower specification limit. However, in some situations, the average is of greater concern.

Take light bulbs for example. A key characteristic is the life expectancy of the bulb. Suppose 20% of brand X light bulbs burn out in less than 800 hours of continuous use. They have an average life of 2000 hours. In contrast only 5% of brand Y light bulbs burn out in less than 800 hours. They have an average life of 1500 hours. Which is the better brand? For most customers, brand X is preferable. It has a longer average life. If one immediately replaces burned out light bulbs, one will use 33% fewer bulbs. In this case, one should inspect for the average rather than percent below 800 hours. If, however, one had a preventive replacement program where bulbs where routinely replaced after 800 hours, brand Y is preferable. It has fewer failures before 800 hours. In this case, one is interested in inspecting for the percent defective.

Special variables sampling plans are available for the averages, mean time between failures, and standard deviations. The OC curves for these variables sampling plans are expressed in terms of the parameter of interest rather than the percent defective. Figure 9.5 (page 201) shows an example of an OC curve for the average. When selecting such sampling plans, instead of specifying the AQL and LTPD in terms of percent defective, one gives values of the parameter being inspected for. In the light bulb example, one might specify an AQL of 1500 hours and LTPD of 1200 hours. Such sampling plans are not covered in this book. Further information can be found in the references at the end of Chapter 1.

A special case of sampling for parameters is bulk sampling. Bulk sampling is performed on continuos lots, i.e., lots that do not consist of units of product. Examples include tanks of solution, batches of cement and plastic blends. One might be inspecting for concentration, particle size, or viscosity. Rather than selecting multiple samples and averaging their results, one can physically blend the samples and test the results once. In the presence of measurement variation, the composite sample can be split and tested several times. Further details are available in Schilling (1982).

9.9   Measurement Error

Variables sampling plans require an estimate of the variation. For known standard deviation plans this is done just once. If the standard deviation is unknown, an estimate of the variation is made for each lot. This estimate of the variation includes both process variation and measurement variation. The variables sampling plans given in Tables 9.6 through 9.9 all assume the measurement variation is small compared to the process variation so that it can be safely ignored. What if the measurement variation is not so small?

The program VARIABLE has an option allowing one to specify that an adjustment be made for measurement variation. This adjustment can only be performed when the standard deviation is known. If you request a measurement adjustment, the program will ask for the percent of the variation due to measurement variation. Let S_t be the estimate of the standard deviation. Both process and measurement variation are included in S_t. Let S_m be an independent estimate of the measurement variation. Then percent variation due to measurement is calculated as follows:

$$\text{Percent variation due to measurement} \;=\; 100 \times \frac{S_m^2}{S_t^2}$$

Screen 9.8: Requesting Measurement Adjustments (VARIABLE)

```
****************************************************************************
***                                                                      ***
***            VARIABLE Option 1 - Enter Variables Sampling Plan         ***
***                                                                      ***
****************************************************************************

    Enter the sample size "n" and constant "k" for the desired variables
    sampling plan.  One must specify whether the standard deviation is
    known or unknown.  If the standard deviation is known, one must also
    specify 1 or 2 spec limits.  In the case of two spec limits, the
    Cp must be entered.  Optionally one can adjust for measurment variation.
    This requires entering the percentage of the total variation resulting.
    from measurement variation.  When the standard deviation is unknown,
    a single spec limit is assumed.

ENTER SAMPLE SIZE     (n)                    --> 26
ENTER CONSTANT        (k)                    --> 2
STANDARD DEVIATION KNOWN? ("y" or "n")       --> y
MEASUREMENT ADJUSTMENT? ("y" or "n")         --> y
PERCENT VARIATION DUE TO MEASUREMENT?        --> 20
NUMBER OF SPEC. LIMITS? (1 or 2)             --> 2
ENTER VALUE OF Cp.                           --> 1.41
```

Screen 9.8 shows the input of a variables sampling plan with a measurement adjustment requested. The measurement variation is estimated to be 20% of the total variation. Screen 9.9 shows the summary statistics for this sampling plan with measurement adjustment. The summary statistics without measurement adjustment were shown in Screen 9.4 (page 200). The measurement adjustment decreases the AQL from 1.01% to 0.470% and the LTPD from 4.02% to 2.53%. The higher the percentage of measurement variation, the more likely lots of all qualities are of being rejected.

Screen 9.9: Summary Statistics with Measurement Adjustment (VARIABLE)

```
                        SUMMARY STATISTICS


    Current Sampling Plan:  n =          26,   k =        2.0000000
                            Standard Deviation Known
                            % Measurement Variation =  20.0000000
                            2 Spec Limits.        Cp =     1.41000


                                               Process
                  Percentile      Symbol    Percent Defective
                  __________      ______    _________________

                     95.0%        AQL              .4704952
                     90.0%                         .5916536
                     50.0%        IQ              1.2673440
                     10.0%        LTPD            2.5287690
                      5.0%                        3.0368330


       AOQL =          .6609796%  (Assumes 100% inspection of rejected
                                   lots with all defectives being found
                                   and repaired plus a large lot size.)

  PRESS ENTER KEY TO CONTINUE
```

Measurement adjustments can also be requested when selecting variables sampling plans under option 6. For a known standard deviation with two specification limits, Screen 9.10 shows the input required to select a variables sampling plan with an AQL of 1% and LTPD of 10%. It is assumed 20% of the variation is due to measurement. The selected sampling plan is n=10 and k=1.56. Screen 9.7 (page 205) shows the selection of the equivalent plan without a measurement adjustment. The selected sampling plan was n=8 and k=1.745. As a result of the measurement adjustment, the sample size increases from 8 to 10. As the percentage of variation due to measurement increases, the number of samples required to achieve the same level of protection also increases. This is a result of it becoming increasingly harder to estimate process performance in the presence of measurement variation.

Screen 9.10: Selection of Plan with Measurement Adjustment (VARIABLE)

```
***************************************************************************
***                                                                   ***
***      VARIABLE Option 6 - Select Sampling Plan Based on AQL and LTPD ***
***                                                                   ***
***************************************************************************

    This option selects the variables sampling plan minimizing the sample
    size from among those satisfying the following set of conditions:

        Actual AQL is greater than or equal to the specified AQL
        Actual LTPD is less than or equal to the specified LTPD

    One must specify whether the standard deviation is known or unknown.
    If the standard deviation is known, one must also specify 1 or 2 spec
    limits.  In the case of two spec limits, the Cp must be entered.
    Optionally one can adjust for measurement variation.  This requires
    entering the percentage of the total variation resulting from
    measurement variation.  When the standard deviation is unknown,
    a single spec limit is assumed.

    ENTER AQL  AS PERCENT DEFECTIVE           --> 1
    ENTER LTPD AS PERCENT DEFECTIVE           --> 10
    STANDARD DEVIATION KNOWN? ("y" or "n")    --> y
    MEASUREMENT ADJUSTMENT? ("y" or "n")      --> y
    PERCENT VARIATION DUE TO MEASUREMENT?     --> 20
    NUMBER OF SPEC. LIMITS? (1 or 2)          --> 2
    ENTER VALUE OF Cp.                        --> 1.5
    USE .95 & .1 FOR AQL-LTPD? ("y" or "n")   --> y
```

Exercises

(9.8) Select a variables sampling plan with an AQL of 1% and LTPD of 5% when the standard deviation is known. Assume there is only one specification limit and that the measurement variation is 20% of the total. What are the actual AQL and LTPD of this sampling plan?

(9.9) Using the sampling plan selected in the previous exercise, how do the AQL and LTPD change if the measurement variation increases to 30% of the total? How about if the measurement variation decreases to 10% of the total?

9.10 Mil-Std-414

Mil-Std-414 is the equivalent of Mil-Std-105E for variables sampling plans. The two standards have many similarities. Both have normal, reduced, and tightened sampling plans along with a set of similar switching rules. Both index the sampling plans by sample size letter code and AQL. Both include a table for determining the sample size letter code based on the lot size and level of inspection. Both provide information on the OC curves, percentiles, and AOQLs.

Despite the many similarities, there is one important difference. These two standards do not use the same level of inspections and corresponding sample size letter codes. This makes it difficult to use Mil-Std-414 to find matching variables sampling plans. To find matching variables sampling plans, one must search through the plots of OC curves and tables of percentiles. Tables 9.8 and 9.9 are better suited for determining matching variables sampling plans.

Mil-Std-414 provides sampling plans for one and two specification limits with known and unknown standard deviations. Two types of variables sampling plans are given: Form 1 and form 2. Form 1 variables sampling plans are described using the sample size n and parameter k. This is the type of variables sampling plan covered in this book. Form 2 sampling plans are described using the sample size n and parameter m. Form 2 variables sampling plans first estimate the percent defective using tables similar to Table 9.1. If two specification limits exist, estimates are made for each specification limit and added together. If this estimate exceeds m, the lot is then rejected. The parameter m is called the maximum allowable percent defective. Form 2 variables sampling plans are given for all situations while form 1 variables sampling plans are only given for the case of a single specification limit.

Mil-Std-414 also provides variables sampling plans for unknown variation based on the average range. These variables sampling plans avoid having to calculate the standard deviation.

9.11   Selecting the Type of Sampling Plan

One has a choice between single sampling plans, double sampling plans, QSSs, and variables sampling plans. QSSs can consist of two single sampling plans or of a double and single sampling plan. Variables sampling plans exist for known and unknown standard deviations. How does one choose between them? One consideration is the number of units that must be inspected. Other considerations include applicability and administrative complexity.

Suppose one wants a sampling plan with an AQL of 1% and LTPD of 4%. Table 9.10 gives corresponding sampling plans of each of these types. The sampling plans are ordered based on their ASN at 1% defective. They are in decreasing order.

Table 9.10: Different Sampling Plans With AQL=1% and LTPD=4%

Type	Parameters	ASN(1%)
Single	n=200, a=4	200.0
Double	n1=110, a1=1, r1=4, n2=145, a2=5	150.0
QSS (single, single)	n_r=120, a_r=3, n_t=165, a_t=3, c=1	122.7
QSS (double, single)	$n1_r$=42, $a1_r$=0, $r1_r$=4, $n2_r$=100, $a2_r$=3, n_t=170, a_t=3, c=2	81.4
Variables Unknown σ	n=80, k=2.01	80.0
Variables Known σ	n=26, k=2.00	26.0

The variables sampling plan with known standard deviation requires the smallest number of units. However, this sampling plan is not always applicable. Its use requires that the inspection of a single characteristic that can be converted to a measurement and that the standard deviation of this measurement is constant and known.

Next the variables sampling plan with unknown standard deviation and the QSS consisting of a double and single sampling plan essentially tie. Both suffer degradation of the protection under certain conditions. The QSS has reduced protection for the first lot following a sudden change in quality. The variables sampling plan has reduced protection for short runs of poor quality and is sensitive to the normality assumption. In addition, the QSS is only applicable to the inspection of a series of lots. When applicable, the QSS is generally the better choice.

Only the top two choices are always applicable and require no assumptions. They also result in the largest number of units inspected. The double sampling plan requires significantly fewer units than the single sampling plan. When a single sample is required, either the single sampling plan or QSS consisting of two single sampling plans must be used. The QSS can only be used if a series of lots is to be inspected.

9.12   Summary

When the standard deviation is known, variables sampling plans can reduce the ASN below that of any other type of sampling plan. However, variables sampling plans require certain assumptions. They can only be applied to a single characteristic that can be converted to a measurement that is normally distributed. For the known standard deviation plans, the standard deviation must also be constant and an accurate estimate obtained. Violations of these assumptions can lead to dramatic changes in the protection provided.

Variables sampling plans should not be used when isolated periods of highly defective product can occur in the middle of a lot or when outliers are present. When periods of highly defective product can occur at the beginning or end of the lot, samples should be stratified. Normally, it is assumed that the measurement variation is negligible. A procedure is provided in the event this is not the case. When the standard deviation is known, acceptance sampling can be combined with control charting using acceptance control charts.

The variables sampling plan used depends on whether the standard deviation is known and on the number of specification limits. Tables are given for a single specification limit. These plans can frequently be applied to the case of double specification limits. In some situations, variables sampling plans for a parameter are more appropriate than variables sampling plans for percent defective.

References

Anderson, Virgil A. (1963). *Design of Experiments.* Marcel Dekker, New York, New York.

Freund, R. A. (1957). "Acceptance Control Charts." Industrial Quality Control, Vol. 14, No. 4, pp 13-23.

Mil-Std-414. (1989). "Sampling Procedures and Tables for Inspection By Variables," U.S. Government Printing Office, Washington, D.C.

Schilling, Edward G. (1982). *Acceptance Sampling in Quality Control.* Marcel Dekker, New York, New York.

Appendix

A

Software
Instructions

The enclosed Sampling Programs diskette contains six programs:

SINGLE: Evaluates and selects single sampling plans. (Chapters 2 and 3)

VALUE: Evaluates and selects single sampling plans to maximize customer value. (Chapter 4)

DOUBLE: Evaluates and selects double sampling plans. (Chapter 6)

QSS: Evaluates and selects quick switching systems. (Chapter 7)

VARIABLE: Evaluates and selects variable sampling plans. (Chapter 9)

PROB: Calculates probabilities from the binomial, hypergeometric, Poisson, normal, and beta distributions. (Chapters 4 and 9, Appendix B)

The chapters and appendices listed after each program describe how to use that program.

The enclosed diskette contains versions of these programs that run under DOS® on an IBM® PC or compatible. This appendix describes how to install and run these programs under DOS. Diskettes containing versions of these programs that run on other machines are available. They come with additional documentation.

A.1 Installation

The programs can be run directly from the enclosed diskette without any installation or setup. This allows you to carry the diskette around with you and run the programs on any available machine so long as you do not violation the license agreement. The programs can also be installed on the hard disk of your PC. Installation is as simple as creating a new subdirectory on your hard disk and copying the programs into it. The rest of this section describes this process in detail.

Start by booting the machine up. This generally requires nothing more than turning on the power. Occasionally, a boot diskette must be put in the A: disk drive. When there is more than one disk drive, the A: drive is the top or left most drive. If the machine boots into DOS you will see a DOS prompt similar to ">", "C:\>" or "C:\DOS>".

If you do not get a DOS prompt, you will see either a DOS shell, a menu, or one of the graphical environments such as Windows[®] or OS2[®]. You must either exit the current environment or start a DOS session under the current environment. If you are in Windows or OS2, double click the left mouse button while pointing to the icon for a DOS session. If you see a DOS shell or menu, you will have to check the documentation that comes with that software for further details.

Once you have the DOS prompt, you must select a disk drive and check that there is sufficient room on it for the programs. To select a drive, type the letter of the drive followed by a colon and press the enter key. For example:

C: 

The A: and B: drives are always floppy drives. The C: drive is generally the first hard disk drive. There may also be other hard drives such as D: and E:.

After selecting a drive, you should check that it contains sufficient space for the programs. This can be done by typing:

CHKDSK 

One of the lines displayed will give the number of bytes available on the disk. This line looks like: "16066560 bytes available on disk." If the number of bytes is greater than 720000, you have sufficient disk space. Otherwise, select a different hard drive or free up some space by deleting unused files.

Once a disk drive with enough memory is found, create a new subdirectory on it for the programs. Placing the programs in a separate subdirectory makes it easier to delete or update them and reduces the chance that they will be accidentally erased. To create a subdirectory named SAMPLING, type:

    MD SAMPLING   (↵   Enter)

Next select the newly created directory by typing:

    CD SAMPLING   (↵   Enter)

You can choose any name you want for the subdirectory.

The last step is to copy the programs on the Sampling Programs diskette into this subdirectory. Assuming the diskette is in the A: drive, type:

    COPY A:*.*   (↵   Enter)

If the diskette is in the B: drive, change the A: to a B:. The installation is now complete. If the enclosed diskette does not fit into any of your floppy drives, find a machine with both types of disk drives and copy the files from the enclosed diskette onto a diskette that fits in your machine.

A.2   Running the Programs

The programs can be run directly from the enclosed diskette. To run a program you must:

1. Insert the Sampling Programs diskette into one of the floppy disk drives.

2. Select that disk drive as the default disk drive.

3. Type the name of the program and press the enter key.

Floppy disk drives are generally the A: and B: drive. Assuming the diskette is in the A: drive, select that drive by typing:

    A: (↵   Enter)

Next type the name of the program and press the enter key. To run the program SINGLE, type:

SINGLE   ⌐⌐ **Enter**

The main menu of the program should then appear.

To run the programs once they have been installed on a hard drive, you must:

1.   Select the hard drive as the default disk drive.

2.   Select the subdirectory containing the programs.

3.   Type the name of the program and press the enter key.

Assuming the hard drive is the C: drive, type:

C: ⌐⌐ **Enter**

Next select the subdirectory containing the programs. Assuming the subdirectory was named SAMPLING, type:

CD \SAMPLING   ⌐⌐ **Enter**

The last step is to type the name of the program and press the enter key. To run the program SINGLE, type:

SINGLE   ⌐⌐ **Enter**

The main menu of the program should then appear.

A.3   Printing and Capturing Screens

If you have a printer attached to your PC, you can print the screen at any time by holding down the shift key and pressing the "Print Scrn" key. Plots or tables can be printed in this manner. This works even if running a DOS session under Windows or OS2.

If you are running a DOS session under Windows or OS2, you can copy an image of the screen to the clipboard. This image can then be pasted into one of the word processors for editing and printing. To copy the screen to the clipboard under Windows, click on the system bar, select the Edit menu item, select the Mark menu item, drag the mouse over the entire image, and press the right mouse button. Further details can be found in the Windows and OS2 manuals.

B

Probability of Acceptance

Appendix B gives the formulas used to calculate the probability of acceptance for single sampling plans, double sampling plans, quick switching systems and variables sampling plans.

B.1   Single Sampling Plans

Single sampling plans have two parameters:

n   =   sample size
a   =   accept number

Either the number of defects or defectives is tallied. Let:

x   =   number of defects or defectives in the sample

Lots where $x \leq a$ are accepted.

Calculating the probability of acceptance requires the density and distribution functions of x. The density function, denoted $f(x \mid n,q)$, is:

$f(x \mid n,q)$ =   probability of exactly x defects or defectives in a sample of size n from a process or lot of quality q.

Using $f(x \mid n,q)$, the distribution function of x can be calculated. The distribution function, denoted $F(x \mid n,q)$, is:

$F(x \mid n,q)$ =   probability of x or fewer defects or defectives in a sample of size n for a process or lot of quality q

= $f(0 \mid n,q) + f(1 \mid n,q) + f(2 \mid n,q) + \cdots + f(x \mid n,q)$

The probability of acceptance, denoted $P_a(q)$, is calculated using the distribution function.   For a single sampling plan, $P_a(q)$ is the probability that x ≤ a, i.e.:

$$P_a(q) \quad = \quad F(a \mid n, q)$$

There are different formulas for $f(x \mid n, q)$ depending on the type of OC curve being calculated. Section 2.3 describes the four types of OC curves. The next four sections are devoted to these four types.

B.2   Type-B OC Curves for Defectives

Type-B OC curves plot the probability of acceptance against the process quality. In the case of defectives, process quality is expressed as a fraction or percent defective. Let:

$$f \quad = \quad \text{process fraction defective}$$

Then the binomial distribution describes the behavior of the number of defective units found in a sample. The binomial distribution has density function $b(x \mid n, f)$ where:

$$b(x \mid n, f) \quad = \quad \text{probability the sample contains exactly x defective units}$$

$$= \quad \binom{n}{x} \, f^x \, (1-f)^{n-x}$$

Calculation of $b(x \mid n, f)$ requires:

$$\binom{n}{x} \;=\; \frac{n!}{x! \; (n-x)!} \qquad \text{and} \qquad z! \;=\; \begin{cases} z(z-1)(z-2)\cdots 1 & z \geq 1 \\ 1 & z = 0 \end{cases}$$

Suppose one takes a sample of 50 units from a process running 1% defective. The process fraction defective is 0.01. The probability the sample contains one defective unit. is $b(1 \mid 50, 0.01)$ which is calculated as follows:

$$b(1 \mid 50, 0.01) \;=\; \binom{50}{1} \, (0.01)^1 \, (1 - 0.01)^{50-1}$$

$$= \; \frac{50!}{1! \; 49!} \, (0.01)^1 \, (0.99)^{49}$$

$$= \; \frac{50 \times 49 \times 48 \times \cdots \times 1}{1 \times 49 \times 48 \times 47 \times \cdots \times 1} \, (0.01)^1 \, (0.99)^{49}$$

$$= \; 50 \, (0.01) \, (0.99)^{49} \qquad = \; 0.3056$$

The distribution function for the binomial is denoted $B(x|n,f)$ where:

$$B(x|n,f) \quad = \quad \text{probability the sample contains x or fewer defectives}$$

$$= \quad b(0|n,f) \; + \; b(1|n,f) \; + \; \cdots \; + \; b(x|n,f)$$

Calculations using these formulas are lengthy. Program PROB on the Sampling Programs diskette can perform these calculations for you. Start the program per the instructions in Appendix A. Starting the program results in Screen B.1 being displayed.

Screen B.1: Menu for Selecting Distribution (PROB)

```
*****************************************************************************
*                            *                                             *
*                            *    Copyright (C) 1992 Taylor Enterprises    *
*      Program PROB          *       P.O. Box 820, Lake Villa, IL 60046    *
*                            *               (708) 356-1074                *
*                            *                                             *
*****************************************************************************


                        Probability Distributions
                        _________________________


                    (1)  Binomial
                    (2)  Hypergeometric
                    (3)  Poisson
                    (4)  Normal
                    (5)  Beta




ENTER NUMBER OF OPTION OR "q" TO QUIT   -->
```

To select the binomial distribution, type "1" and press the enter key. The program will then request values for x, n and f. Type each value followed by the enter key. Screen B.2 shows the input required to calculate the probability of a single defective in a sample of 50 units from a process running 1% defective (0.01 fraction defective). The values entered by the user are preceded by the prompt "-->". The probability of 1 defective, $b(1|50,0.01)$, is 0.3056. This agrees with the previously calculated value.

Using the binomial distribution, the Type-B OC curve for defectives of a single sampling plan is

$$P_a(p) \quad = \quad B(a|n,p/100)$$

where p is the process percent defective. The probability of the single sampling plan n=50 and a=1 accepting a 1% defective lot is $B(1|50,0.01)$. Screen B.2 shows that $B(1|50,0.01) = 0.911$.

Screen B.2: Screen for Binomial Probabilities (PROB)

```
********************************************************************************
***                                                                        ***
***                  PROB Option 1 - Binomial Distribution                 ***
***                                                                        ***
********************************************************************************

   Both b(x|n,f) and B(x|n,f) are calculated for the binomial distribution.
   It is assumed that a sample of size n has been selected from a process
   with fraction defective f.  Then:

      b(x|n,f)  =  probability of exactly x defective units in the sample
      B(x|n,f)  =  probability of x or fewer defective units in the sample


ENTER x --> 1
ENTER n --> 50
ENTER f --> .01


   Results:   b(x|n,f)  =    .30555862
              B(x|n,f)  =    .91056469



ENTER "r" TO REPEAT, "m" FOR MAIN MENU, OR "q" TO QUIT -->
```

B.3   Type-A OC Curves for Defectives

Type-A OC curves plot the probability of acceptance against lot
quality. Let:

N = lot size

X = number of defectives in the lot

Then the hypergeometric distribution describes how the number of
defective units in the sample behaves. The hypergeometric
distribution has density function $h(x\,|\,n,X,N)$ where:

$$h(x\,|\,n,X,N) = \text{probability the sample contains exactly } x \text{ defective units}$$

$$= \frac{\binom{n}{x}\binom{N-n}{X-x}}{\binom{N}{X}}$$

Suppose one takes a sample of 50 units from a lot of 1000 units
containing 10 defectives. This lot is 1% defective. Then
$h(1\,|\,50,10,1000)$ is the probability the sample contains one defective.
It is calculated as follows:

$$h(1\,|\,50,10,1000) \;=\; \frac{\binom{50}{1}\binom{1000-50}{10-1}}{\binom{1000}{10}} \;=\; \frac{\left(\dfrac{50!}{1!\;49!}\right)\left(\dfrac{950!}{9!\;941!}\right)}{\left(\dfrac{1000!}{10!\;990!}\right)}$$

$$=\; \frac{\left(\dfrac{50}{1}\right)\left(\dfrac{950\times949\times948\times\cdots\times942}{9\times8\times7\times\cdots\times1}\right)}{\left(\dfrac{1000\times999\times998\times\cdots\times991}{10\times9\times8\times\cdots\times1}\right)}$$

$$=\; 50\,\times\,10\,\times\,\frac{950\times949\times948\times\cdots\times942}{1000\times999\times998\times\cdots\times991}$$

$$=\; 0.3174$$

The distribution function for the hypergeometric is denoted $H(x\,|\,n,X,N)$ where:

$$H(x\,|\,n,X,N) \;=\; \text{probability the sample contains x or fewer defectives}$$

$$=\; h(0\,|\,n,X,N) \,+\, h(1\,|\,n,X,N) \,+\, \cdots \,+\, h(x\,|\,n,X,N)$$

Values of $h(x\,|\,n,X,N)$ and $H(x\,|\,n,X,N)$ can be obtained by running the program PROB. Start the program as before. Once Screen B.1 appears, select the hypergeometric distribution by typing "2" and pressing the enter key. The program will ask for x, n, X, and N. Type each value followed by the enter key. Screen B.3 shows the input required to calculate the probability of a single defective in a sample of 50 units from a lot of 1000 units containing 10 defectives. The probability of one defective, $h(1\,|\,50,10,1000)$, is 0.3174. This agrees with the previously calculated value.

Using the hypergeometric distribution, the Type-A OC curve for defectives of a single sampling plan is

$$P_a(p) \;=\; H(a\,|\,n,X,N)$$

where p is the lot percent defective making

$$p \;=\; 100\,(^X\!/\!N)$$

The probability of the single sampling plan n=50 and a=1 accepting a lot consisting of 1000 units, 10 of which are defective is $H(1\,|\,50,10,1000)$. Screen B.3 shows that $H(1\,|\,50,10,1000) = 0.915$.

Screen B.3: Screen for Hypergeometric Probabilities (PROB)

```
********************************************************************************
***                                                                        ***
***               PROB Option 2 - Hypergeometric Distribution              ***
***                                                                        ***
********************************************************************************

   Both h(x|n,X,N) and H(x|n,X,N) are calculated for the hypergeometric
   distribution.  It is assumed that a sample of size n has been selected
   from a lot of N units containing a total of X defective units.  Then:

      h(x|n,X,N)  =  probability of exactly x defective units in the sample
      H(x|n,X,N)  =  probability of x or fewer defective units in the sample


ENTER x --> 1
ENTER n --> 50
ENTER X --> 10
ENTER N --> 1000


   Results:    h(x|n,X,N)  =      .31738113
               H(x|n,X,N)  =      .91469243


ENTER "r" TO REPEAT, "m" FOR MAIN MENU, OR "q" TO QUIT -->
```

B.4   Type-B OC Curves for Defects

Type-B OC curves plot the probability of acceptance against the
process quality. In the case of defects, process quality is expressed in
terms of the average number of defects per unit.   Let:

$$r \quad = \quad \text{average number of defects per unit for the process}$$

Then the Poisson distribution describes the behavior of the number of
defects in a sample. The Poisson distribution has density function
$p(x|n,r)$ where:

$$p(x|n,r) \quad = \quad \text{probability the sample contains exactly x defects}$$

$$= \quad \frac{(nr)^x \, e^{-nr}}{x!}$$

Suppose one takes a sample of 50 units from a process that is
averaging 0.01 defects per unit. The probability the sample contains
one defect is $p(1|50,0.01)$ which is calculated as follows:

$$p(1|50,0.01) \quad = \quad \frac{(50 \times 0.01)^1 \, e^{-50 \times 0.01}}{1!} \quad = \quad 0.3033$$

The distribution function for the Poisson is denoted $P(x|n,r)$ where:

$$P(x|n,r) \quad = \quad \text{probability the sample contains x or fewer defects}$$

$$= \quad p(0|n,r) \; + \; p(1|n,r) \; + \; \cdots \; + \; p(x|n,r)$$

For the Poisson distribution, the sample size n can take non integer values.   Suppose one takes a sample of 0.5 cubic feet of cement and inspects it for foreign objects, clumps, and oversize stones. Assume the process and materials for producing the cement average 10 of these defects per cubic foot. Then p(5|0.5, 10) is the probability of 5 defects in the sample.

Values of p(x|n,r) and P(x|n,r) can be obtained by running the program PROB. Start the program as before and select the Poisson distribution by typing "3" and pressing the enter key. The program will ask for x, n and r. Screen B.4 shows input required to calculate the probability of a single defect in a sample of fifty units from a process averaging 0.01 defects per unit. The probability of one defective, p(1|50,0.01), is 0.3033. This agrees with the previously calculated value.

Screen B.4: Screen for Poisson Probabilities (PROB)

```
**********************************************************************************
***                                                                          ***
***                 PROB Option 3 - Poisson Distribution                     ***
***                                                                          ***
**********************************************************************************

   Both p(x|n,r) and P(x|n,r) are calculated for the Poisson distribution.
   It is assumed that a sample of size n has been selected from a lot that
   averages r defects per unit.  n does not have to be an integer.  Then:

      p(x|n,r)  =  probability of exactly x defects in the sample
      P(x|n,r)  =  probability of x or fewer defects in the sample


ENTER x --> 1
ENTER n --> 50
ENTER r --> .01


   Results:   p(x|n,r)  =     .30326533
              P(x|n,r)  =     .90979599



ENTER "r" TO REPEAT, "m" FOR MAIN MENU, OR "q" TO QUIT -->
```

Using the Poisson distribution, the Type-B OC curve for defectives of a single sampling plan is

$$P_a(r) \;=\; P(a|n,r)$$

where r is the average number of defects per unit for process. The probability of the single sampling plan n=50 and a=1 accepting a process averaging 0.01 defects per unit is P(1|50,0.01). Screen B.4 shows that P(1|50,0.01) = 0.910.

B.5   Type-A OC Curves for Defects

Type-A OC curves plot the probability of acceptance against lot quality. Let:

$$N \;=\; \text{lot size}$$
$$X \;=\; \text{number of defects in lot}$$

Then the density function for the number of defects in the sample, denoted $m(x\,|\,n,X,N)$, can be expressed in terms of the binomial density function as follows:

$$m(x\,|\,n,X,N) \;=\; \text{probability the sample contains exactly x defects}$$

$$=\; b(x\,|\,X,n/N)$$

Suppose one takes a sample of 50 units from a lot of 1000 units containing 10 defects. Then $m(1\,|\,50,10,1000)$ is the probability the sample contains one defect. It is calculated as follows:

$$m(1\,|\,50,10,1000) \;=\; b(1\,|\,10,50/1000)$$

$$=\; \binom{10}{1} \left(50/1000\right)^{1} \left(1-50/1000\right)^{10-1}$$

$$=\; \frac{10!}{1!\;9!}\,(0.05)^{1}\,(0.95)^{9} \qquad =\; 0.3151$$

The distribution function for the number of defects is denoted $M(x\,|\,n,X,N)$ where:

$$M(x\,|\,n,X,N) \;=\; \text{probability the sample contains x or fewer defects}$$

$$=\; B(x\,|\,X,n/N)$$

The binomial distribution applies since each of the X defects has probability n/N of being in the sample. Further details are in Taylor (1992a). The sample size and lot size can take non integer values.

Values for $b(x\,|\,X,n/N)$ and $B(x\,|\,X,n/N)$ can be obtained by running the program PROB. Start the program as before and select the binomial distribution by typing "1" and pressing the enter key. The program will ask for x, n and f. When n is requested, type the value of X. When f is requested, type the value of n/N. Screen B.5 shows the input required to calculate the probability of a single defect in a sample of fifty units from a lot of 1000 units containing 10 defects.

The probability of one defect, b(1 | 10, $^{50}\!/_{1000}$), is 0.3151. This agrees with the previously calculated value.

Screen B.5: Probabilities for Type-A OC Curves for Defects (PROB)

```
****************************************************************************
***                                                                      ***
***                  PROB Option 1 - Binomial Distribution               ***
***                                                                      ***
****************************************************************************

    Both b(x|n,f) and B(x|n,f) are calculated for the binomial distribution.
    It is assumed that a sample of size n has been selected from a process
    with fraction defective f.  Then:

       b(x|n,f)  =  probability of exactly x defective units in the sample
       B(x|n,f)  =  probability of x or fewer defective units in the sample


ENTER x --> 1
ENTER n --> 10
ENTER f --> .05


   Results:   b(x|n,f)  =     .31512471
              B(x|n,f)  =     .91386164



ENTER "r" TO REPEAT, "m" FOR MAIN MENU, OR "q" TO QUIT -->
```

Using these probabilities, the probability of acceptance for a single sampling plan can be calculated as follows:

$$P_a(r) = B(a \mid X, {}^n\!/_N)$$

where the average number of defects per unit in the lot is:

$$r = {}^X\!/_N$$

The probability of the single sampling plan n=50 and a=1 accepting a lot consisting of 1000 units containing 10 defects is $B(1 \mid 10, {}^{50}\!/_{1000})$. Screen B.5 shows that $B(1 \mid 10, {}^{50}\!/_{1000}) = 0.914$. Table B.1 below summarizes the formulas for probability of acceptance for the different types of OC curves.

Table B.1: Probability of Acceptance for Single Sampling Plans

Tally?	Probability of Acceptance
Type-B for Defectives	$P_a(p) = B(a \mid n, {}^p\!/_{100})$
Type-A for Defectives	$P_a(p) = H(a \mid n, X, N)$ $p = 100\,{}^X\!/_N$
Type-B for Defects	$P_a(r) = P(a \mid n, r)$
Type-A for Defects	$P_a(r) = B(a \mid X, {}^n\!/_N)$ $r = {}^X\!/_N$

B.6   Double Sampling Plans

Double sampling plans have five parameters: namely n_1, a_1, r_1, n_2 and a_2. Either the number of defects or defectives is tallied. Let:

$$x_1 = \text{number of defects or defectives in the first sample}$$
$$x_2 = \text{number of defects or defectives in the second sample}$$

The lot is accepted if x_1 is less than or equal to a_1 or if x_1 is more than a_1 but less than r_1 and $x_1 + x_2$ is less than or equal to a_2. For a double sampling plan, the probability of acceptance, denoted $P_a(q)$, is calculated as follows:

$$P_a(p) \;=\; \Pr\{x_1 \le a_1\} + \sum_{i=a_1+1}^{r_1-1} \left[\Pr\{x_1 = a_1\} \times \Pr\{x_1 + x_2 \le a_2\}\right]$$

The term $\Pr\{x_1 \le a_1\}$ is read as the probability that x_1 is less than or equal to a_1.

For Type-B OC curves for defectives, both x_1 and x_2 are binomially distributed so x_1 has density $b(x_1 \mid n_1, p)$ and x_2 has density $b(x_2 \mid n_2, p)$. In this case the probability of acceptance is

$$P_a(p) \;=\; B(a_1 \mid n_1, p) + \sum_{i=a_1+1}^{r_1-1} \left[b(i \mid n_1, p) \times B(a_2 - i \mid n_2, p)\right]$$

Similar formulas exist for the other types of OC curves. These formulas are:

$$P_a(r) \;=\; P(a_1 \mid n_1, r) + \sum_{i=a_1+1}^{r_1-1} \left[p(i \mid n_1, r) \times P(a_2 - i \mid n_2, r)\right]$$

$$P_a\!\left(p = 100\tfrac{X}{N}\right) = H(a_1 \mid n_1, X, N) + \sum_{i=a_1+1}^{r_1-1} \left[h(i \mid n_1, X, N) \times H(a_2 - i \mid n_2, X - i, N - n_1)\right]$$

$$P_a\!\left(r = \tfrac{X}{N}\right) \;=\; B\!\left(a_1 \mid X, \tfrac{n_1}{N}\right) + \sum_{i=a_1+1}^{r_1-1} \left[b\!\left(i \mid X, \tfrac{n_1}{N}\right) \times B\!\left(a_2 - i \mid X - i, \tfrac{n_2}{N-n_1}\right)\right]$$

Further information concerning these formulas is available in Guenther (1982) and Taylor (1992a).

B.7   Quick Switching Systems

The probability of acceptance for a QSS is a weighted average of the probabilities of acceptance for the reduced and tightened inspections. The weights are the probabilities of being in reduced and tightened inspection respectively. Let:

$PR_a(p)$ = Probability of acceptance for reduced inspection

$S_{R \to T}(p)$ = Probability switch from reduced to tightened

$PT_a(p)$ = Probability of acceptance for tightened inspection

$S_{T \to R}(p)$ = Probability switch from tightened to reduced

Then the probability of acceptance for a QSS, denoted $P_a(p)$, is

$$P_a(p) \;=\; \frac{S_{T \to R}\, PR_a(p) + S_{R \to T}\, PT_a(p)}{S_{T \to R} + S_{R \to T}}$$

Since QSSs switch from reduced to tightened on rejection: $S_{R \to T}(p) = PR_a(p)$. For tightened to reduced: $S_{T \to R}(p) = F(s_t \mid n_t, p)$. Further details can be found in Taylor (1992b).

B.8   Variables Sampling Plans

Formulas for variables sampling plans can be found in Guenther (1977).

References

Guenther, William C.. (1977). *Sampling Inspection in Statistical Quality Control*. MacMillan, New York.

Taylor, Wayne A. (1992a). "Type-A OC curves for the Poisson Distribution." Submitted for publication.

Taylor, Wayne A. (1992b). "Quick Switching Systems: Part 1 - Evaluation." Submitted for publication.

Average
Sample Number

This appendix gives the formulas used to calculate the average sample number for single sampling plans, double sampling plans, quick switching systems and variable sampling plans.

C.1   Single Sampling Plans

Single sampling plans have two parameters:

    n = sample size

    a = accept number

The ASN depends on whether defects or defectives are tallied and on the method of curtailing. When defectives are tallied, the ASN is a function of the process percent defective, denoted p. When defects are tallied, the ASN is a function of the average number of defects for the process, denoted r. Further, for defects, the ASN depends on whether lots consist of units or are continuous. The ASN is calculated as follows:

$$\text{ASN} \;=\; \text{ASN}_{\text{accept}}(n,a) \;+\; \text{ASN}_{\text{reject}}(n,a)$$

Table C.1 and C.2 gives formulas for $\text{ASN}_{\text{accept}}(n,a)$ and $\text{ASN}_{\text{reject}}(n,a)$. These formulas use the binomial and Poisson density and distribution functions from Appendix B. The derivation of these formulas is given in Guenther (1977) and Guenther (1983). If defectives are tallied and no curtailing is performed:

$$\text{ASN} \;=\; n\, B\!\left(a \,\middle|\, n, \tfrac{p}{100}\right) \;+\; n\left[1 - B\!\left(a \,\middle|\, n, \tfrac{p}{100}\right)\right] \;=\; n$$

Table C.1: Formulas for ASNaccept(n,a)

Tally?	Curtail on Acceptance?	$\mathrm{ASN}_{\mathrm{accept}}(n,a)$
Defectives	no	$n\,B\!\left(a\middle|n,\tfrac{p}{100}\right)$
	yes	$\dfrac{n-a}{1-\tfrac{p}{100}}\,B\!\left(a\middle|n+1,p\right)$
Defects - Lot Contains Units	no	$n\,P\!\left(a\middle|n,r\right)$
Defects - Continuous Lot	no	$n\,P\!\left(a\middle|n,r\right)$

Table C.2: Formulas for $\mathrm{ASN}_{\mathrm{reject}}(n,a)$

Tally?	Curtail on Rejection?	$\mathrm{ASN}_{\mathrm{reject}}(n,a)$
Defectives	no	$n\left[1-B\!\left(a\middle|n,\tfrac{p}{100}\right)\right]$
	yes	$\dfrac{a+1}{\tfrac{p}{100}}\left[1-B\!\left(a+1\middle|n+1,\tfrac{p}{100}\right)\right]$
Defects - Lot Contains Units	no	$n\left[1-P\!\left(a\middle|n,r\right)\right]$
	yes	$1+\displaystyle\sum_{j=1}^{n-1}P\!\left(a\middle|j,r\right)-n\,P\!\left(a\middle|n,r\right)$
Defects - Continuous Lot	no	$n\left[1-P\!\left(a\middle|n,r\right)\right]$
	yes	$\dfrac{a+1}{r}\left[1-P\!\left(a+1\middle|n+1,r\right)\right]$

C.2   Double Sampling Plans

Double sampling plans have five parameters:

$$n1 \quad = \quad \text{first sample size}$$
$$a1 \quad = \quad \text{first accept number}$$
$$r1 \quad = \quad \text{first reject number}$$
$$n2 \quad = \quad \text{second sample size}$$
$$a2 \quad = \quad \text{second accept number}$$

Based on these paramters the ASN is:

$$
\text{ASN} = \begin{aligned}
&\text{ASN}_{\text{accept}}(n1, a1) + \text{ASN}_{\text{reject}}(n1, r1 - 1)\\
&+ \sum_{i=a1+1}^{r1-1}\left[n1 + b(i|n1, p)\big(\text{ASN}_{\text{accept}}(n2, a2 - i) + \text{ASN}_{\text{reject}}(n2, a2 - i)\big)\right]
\end{aligned}
$$

C.3   Quick Switching Systems

For QSSs, the ASN is the weighted average of the ASNs of the reduced and tightened inspections. Using the notation from Section B.7 and letting ASN_r and ASN_t denote the ASNs of the reduced and tightened inspection:

$$
\text{ASN} = \frac{S_{T \to R}\, \text{ASN}_r + S_{R \to T}\, \text{ASN}_t}{S_{T \to R} + S_{R \to T}}
$$

C.4   Variables Sampling Plans

For variables sampling plans, the ASN is always equal to the sample size.

References

Guenther, William C.. (1977). *Sampling Inspection in Statistical Quality Control.* MacMillan, New York.

Guenther, William C.. (1983). "Average Sample Number for Semi-Curtailed Sampling Using the Poisson." Journal of Quality Technology, Vol. 15, No. 3, pp. 126-129.

Chapter 2. Evaluating Single Sampling Plans

(2.1)   Reject the first lot. Accept the second lot.

(2.2)   Number of defects since units do not exist.

(2.3)   Either, since the defect rate is low.

(2.4)   Number of defects, since the defect rate is high.

(2.5)   A Type A OC curve for defectives with N=100 should be plotted. $P_a(1\%) = 0.870$, $P_a(8\%) = 0.314$, $P_a(16\%) = 0.088$.

(2.6)   A Type A OC curve for defectives with N=10,000 should be plotted. $P_a(1\%) = 0.877$, $P_a(8\%) = 0.338$, $P_a(16\%) = 0.104$.

(2.7)   A Type B OC curve for defectives should be plotted. $P_a(1\%) = 0.878$, $P_a(8\%) = 0.338$, $P_a(16\%) = 0.104$.

(2.8)   A Type B OC curve for defects should be plotted. $P_a(0.01) = 0.878$, $P_a(0.08) = 0.353$, $P_a(0.16) = 0.125$.

(2.9)   A Type A OC curve for defects with N=500 should be plotted. $P_a(0.01) = 0.877$, $P_a(0.08) = 0.349$, $P_a(0.16) = 0.122$.

(2.10)   From the Type A OC curve with N=100: AQL = 0.385%, LTPD = 15.2%.

(2.11)   From the Type A OC curve with N=10,000: AQL = 0.394%, LTPD = 16.2%.

(2.12)   From the Type B OC curve: AQL = 0.394%, LTPD = 16.2%.

(2.13)   From the Type A OC curve with N=100: AQL = 0.00385, LTPD = 0.166.

(2.14)   From the Type B OC curve: AQL = 0.00395, LTPD = 0.177.

(2.15)   AOQ(0.005) = 0.00469, AOQ(0.01) = 0.00880,
         AOQ(0.05) = 0.0270.

(2.16)   ASN(1%) = 48.44, ASN(5%) = 32.87.

(2.17)   ASN(1%) = 49.39, ASN(5%) = 49.92.

(2.18)   ASN(0.01) = 48.37, ASN(0.05) = 32.61.

Chapter 3.   Selecting Single Sampling Plans

(3.1)   A Type B OC curve for defectives should be used. Program
         SINGLE gives n=79 and a=1. The closest match in Table 3.3
         is n=80 and a=1.

(3.2)   A Type A OC curve for defectives with N=500 should be used.
         Program SINGLE gives n=73 and a=1.

(3.3)   A Type B OC curve for defects should be used. Program
         SINGLE gives n=78 and a=1. The ratio of LTPD to AQL is
         12.5. Table 3.4 gives an accept number of 1 and a sample size
         range of 77.80 to 88.85. The smallest integer sample size in
         this range is 78.

(3.4)   A Type A OC curve for defects with N = 500 should be used.
         Program SINGLE gives n=74 and a=1.

(3.5)   A Type B OC curve for defects should be used. Specifying a
         continuous lot and a smallest sample size increment of 5,
         program Single gives n=80 and a=1. The ratio of LTPD to
         AQL is 12.5 so Table 3.4 gives an accept number of 1 and a
         sample size range of 77.80 to 88.85. The smallest multiple of
         5 in this range is 80.

(3.6)   Program SINGLE gives n=84 and a=1. The closest match in
         Table 3.3 is n=80 and a=1.

(3.7)   Program SINGLE gives n=84 and a=1. The ratio of AOQL to
         AQL is 2.5. Table 3.5 gives an accept number of 1 and a
         sample size range of 84.00 to 88.85. The smallest integer
         sample size in this range is 84.

(3.8)   Program SINGLE gives n=140 and a=2. The ratio of AOQL to
         AQL is 2.5. Table 3.5 gives an accept number of 1 and a
         sample size range of 84.00 to 88.85. No multiple of 10 falls in
         this range so increment the accept number to 2. The new
         range is 137.10 to 204.42. The smallest multiple of 10 in this
         range is 140.

(3.9)   n=5 and a=0.

(3.10)   n=5 and a=0.

Chapter 4.   Economical Single Sampling Plans

(4.1)   The single sampling plan n=52 and a=7 where rejected lots are discarded. However, the resulting customer value, $2502.95, is only $2.95 above the customer value for the decision release all lots.

(4.2)   Yes for R=$2, since R<P-D=$5. No for R=$6, since R>P-D.

(4.3)   The single sampling plan n=292 and a=1 where rejected lots are 100% inspected. For this plan the customer value is $28,068.29 per lot or $2.81 per unit. This compares to a value of $2.74 per unit when the lot size was 1000 for a potential of saving 7 cents per unit. This savings assumes the distribution of process quality is not effected by the change in lot size.

(4.4)   The customer value for a 100% inspection with U_h=$0.003, S_h=0, E=80% and no rework is $2857.00 per lot. The customer value for the current plan, n=80 and a=0, is $2738.28. The suggestion will save $118.72 per lot, above and beyond repaying implementation costs. The suggestion should be implemented.

(4.5)   When U_p = 0.5% and S_p = 1.5%, the single sampling plan n=56 and a=0 where rejected lots are 100% inspected maximizes the customer value. The customer value for this plan is $2847.79. Under these same conditions the customer value for the current plan n=80, a=0 is $2843.89. The benefit from changing plans is small.

When U_p = 0.5% and S_p = 0.5% the action maximizing the customer value is to release all lots with a customer value of $2750.00. Under these same conditions the customer value for the current plan n=80 and a=0 is $2724.76. The benefit from ceasing inspection is $25 per lot.

(4.6)   Beta (5%|2%, 3%) = 0.880

(4.7)   n=25, m=500, $\sum_{i=1}^{m} d_i$ = 186, $\sum_{i=1}^{m} d_i^2$ = 648, U_p = 1.49%, and S_p = 3.63%.

(4.8)   S.E.(U_p)=0.192% so the true value of U_p should be in the interval (1.11%, 1.87%). S.E.(S_p)=0.261% so the true value of S_p should be in the interval (3.11%, 4.15%).

(4.9) $m=1000, \sum_{i=1}^{m}\left(\dfrac{d_i}{n_i}\right) = 4.592, \sum_{i=1}^{m}\left(\dfrac{d_i}{n_i}\right)^2 = 0.9347, \sum_{i=1}^{m}\left(\dfrac{1}{n_i}\right) = 71.92,$

$U_p = 0.459\%$, and $S_p = 2.51\%$.

(4.10) The plan n=114 and a=0 maximizes the customer value with a value of \$2732.62. This is \$5.66 per lot below that of n=80 and a=0.

Chapter 6.   Double Sampling Plans

(6.2)   Reject in both cases.

(6.3) $P_a(1\%) = 1.00$, $P_a(4\%) = 0.266$, $P_a(8\%) = 0.0412$.

(6.4) $P_a(1\%) = 0.899$, $P_a(4\%) = 0.361$, $P_a(8\%) = 0.0784$.

(6.5) $P_a(1\%) = 0.895$, $P_a(4\%) = 0.368$, $P_a(8\%) = 0.0823$.

(6.6)   As the lot size increases, the OC curve becomes steeper.

(6.7) $P_a(1\%) = 0.895$, $P_a(4\%) = 0.377$, $P_a(8\%) = 0.0926$.

(6.8)   AQL = 1.16, IQ = 2.75, LTPD = 6.05.

(6.9)   AQL = 0.680, IQ = 3.07, LTPD = 7.37.

(6.10)   AQL = 0.650, IQ = 3.11, LTPD = 7.49.

(6.11)   AQLs decrease to Type B AQL and LTPDs increase to Type B LTPD.

(6.12)   AQL = 0.00646, IQ = 0.0315, LTPD = 0.0778, AOQL = 0.0157.

(6.13)   AOQ(1%) = 0.905, AOQ(4%) = 1.73, AOQ(8%) = 1.40.
         The AOQ curve is always higher than the one in Figure 6.3.

(6.14)   ASN(1%) = 38.4, ASN(4%) = 38.6, ASN(8%) = 34.2. The ASN curve is between the ASN curves for no curtailing and full curtailing.

(6.15)   n1=47, a1=0, r1=2, n2=60, a2=1.

(6.16)   n1=45, a1=0, r1=2, n2=162, a2=2.

(6.17)   n1=0.08, a1=1, r1=4, n2=0.14, a2=4.

(6.18)   The single sampling plan satisfying these requirements is n=69 and a=2. It has an AQL of 1.20% and LTPD of 7.53%. The matching double sampling plan (AQL=1%, LTPD=7.53%) minimizing ASN(1%) is n1=33, a1=0, r1=3, n2=51 and a2=2. It has an actual AQL of 1.07% and actual AOQL of 1.83%. These satisfy the initial requirements.

Chapter 7.   Quick Switching Systems

(7.2)   Reject and use tightened inspection for next lot.

(7.3) $P_a(1\%) = 0.858$, $P_a(4\%) = 0.416$, $P_a(8\%) = 0.110$.

(7.4) $P_a(1\%) = 0.878$, $P_a(4\%) = 0.588$, $P_a(8\%) = 0.338$.
They are the same.

(7.5) $P_a(1\%) = 0.740$, $P_a(4\%) = 0.294$, $P_a(8\%) = 0.0820$.
They are the same.

(7.6) $P_a(0.01) = 0.859$, $P_a(0.04) = 0.426$, $P_a(0.08) = 0.123$.

(7.7)   AQL = 0.370, IQ = 3.36, LTPD = 8.27, AOQL = 1.68,
maximum chance of rejection at AQL = 0.105,
rate of convergence at AQL = 89.5%,
maximum chance of acceptance at LTPD = 0.325, and
rate of convergence at LTPD = 67.5%.

(7.8)   AQL = 0.00371, IQ = 0.0342, LTPD = 0.0864, AOQL = 0.0171,
maximum chance of rejection at AQL = 0.106,
rate of convergence at AQL = 89.5%,
maximum chance of acceptance at LTPD = 0.325, and
rate of convergence at LTPD = 67.5%.

(7.9)   AOQ(1%) = 0.913, AOQ(4%) = 1.91, AOQ(8%) = 1.42.
The AOQ curve is always higher than the one in Screen 7.8.

(7.10)   ASN(1%) = 15.4, ASN(4%) = 22.9, ASN(8%) = 28.1.

(7.11)   From the program: $n1_r=25$, $a1_r=0$, $r1_r=2$, $n2_r=55$, $a2_r=1$, $n_t=78$, $a_t=1$ and $s_t=0$. From the table: $n1_r=26$, $a1_r=0$, $r1_r=2$, $n2_r=160$, $a2_r=1$, $n_t=80$, $a_t=1$ and $s_t=0$.

(7.12) $n1_r=24$, $a1_r=0$, $r1_r=2$, $n2_r=75$, $a2_r=1$, $n_t=78$, $a_t=1$ and $s_t=0$.

(7.13) $n1_r=0.03$, $a1_r=0$, $r1_r=3$, $n2_r=0.09$, $a2_r=3$, $n_t=0.11$, $a_t=2$ and $s_t=1$.

(7.14) $n1_r=32$, $a1_r=0$, $r1_r=2$, $n2_r=32$, $a2_r=1$, $n_t=85$, $a_t=1$ and $s_t=0$. The AQL remains fairly constant changing from 0.647% to 0.619%. The LTPD is improved from 7.50% to 4.58%. The ASN at 0.65% increases only slightly changing from 37.4 to 41.3 (no curtailing).

Chapter 8.   Mil-Std-105E

(8.1)   The single sampling plan n=125 and a=1. The sample size letter code is J. One must follow the arrow one row down.

(8.2) P_a(0.4%) = 0.910, AQL = 0.285%, LTPD = 3.08%. It is indexed under an AQL of 0.25% in Table 3.3.

(8.3)   The original single sampling plan has an AQL of 1.03% and LTPD of 6.52%. The matching double sampling plan is n1=38, a1=0, r1=3, n2=90 and a2=3. It has an AQL of 1.14% and LTPD of 6.40%. The ASN(0%) = 38.0, ASN(0.5%) = 53.5 and ASN(1%) = 66.0. The matching QSS is $n1_r$=20, $a1_r$=0, $r1_r$=2, $n2_r$=60, $a2_r$=2, n_t=60, a_t=1 and s_t=0. It has an AQL of 1.07% and LTPD of 6.44%. The ASN(0%) = 20.0, ASN(0.5%) = 25.8 and ASN(1%) = 31.8.

(8.4)   The original single sampling plan has an AQL of 0.256% and LTPD of 10.9%. There is no matching double sampling plan. The QSS n_r=20, a_r=0, n_t=40, a_t=0 and s_t=0 provides improved protection at a similar cost. It has an AQL of 0.244% and LTPD of 6.13%. The ASN(0%) = 20.0, ASN(0.125%) = 20.5 and ASN(0.25%) = 21.0.

Chapter 9.   Variables Sampling Plans

(9.1)   Mean = 51.97, standard deviation = 0.6464.

(9.2)   0.0 + 0.621% = 0.621% defective

(9.3)   Accept if $3.2483 \leq \overline{X} \leq 3.2517$. Accept first lot, reject second.

(9.4)   Accept since 4+1.68×0.23 = 4.39 ≤ 4.6.

(9.5)   AQL = 1.01%, LTPD = 4.02%, AOQL = 1.24%, P_a(2%) = 0.609, AOQ(2%) = 1.35, ASN(2%) = 26.

(9.6)   n = 8 and k = 1.745. It is the same as the double specification limit plan. Accept if $25.0+1.745×2.2 = 28.84 \leq \overline{X}$.

(9.7)   n = 21 and k = 1.761. The sample size more than doubles when the standard deviation is not known. Accept since 25.0+1.761×1.24 = 27.18 ≥ 26.57.

(9.8)   n = 24 and k = 1.745. AQL = 0.999% and LTPD = 4.86%.

(9.9)   When 30%, AQL = 0.644% and LTPD = 3.81%.
When 10%, AQL = 1.41% and LTPD = 5.90%.

Index